Frederick Street

MAUDE BARLOW & ELIZABETH MAY

Frederick Street

Life and Death on Canada's Love Canal

A Phyllis Bruce Book
HarperCollins*PublishersLtd*

FREDERICK STREET: LIFE AND DEATH
ON CANADA'S LOVE CANAL
 For information address HarperCollins Publishers Ltd, 55 Avenue Road, Suite 2900, Toronto, Ontario, Canada M5R 3L2.

www.harpercanada.com

HarperCollins books may be purchased for educational, business, or sales promotional use. For information please write: Special Markets Department, HarperCollins Canada,
55 Avenue Road, Suite 2900, Toronto,
Ontario, Canada M5R 3L2.

First HarperCollins hardcover ed.
ISBN 0-00-200036-9
First HarperCollins trade paper ed.
ISBN 0-00-638529-X

The authors are deeply indebted to the Canada Council for its support in the writing of this book.

Map design, page xii: Gary Hutton

Permission has been granted to use the following excerpts:

Shirley Christmas Kiju Kawi, "Within My Dreams," Sydney: Mukla'qati Press, 1994. By permission of the author.
Leon Dubinsky, "We Rise Again." By permission of the author.
Max MacDonald, "Test of Mettle." By permission of the author.
Kenzie MacNeil, "The Island." By permission of the author.

Canadian Cataloguing in Publication Data

Barlow, Maude
Frederick Street:
life and death on Canada's Love Canal

A Phyllis Bruce book.
ISBN 0-00-200036-9

1. Hazardous waste sites – Health aspects – Nova Scotia – Sydney.
2. Steel industry and trade – Waste disposal – Health aspects – Nova Scotia – Sydney.
I. May, Elizabeth.
II. Title.

TD195.S7B37 2000 363.738'09716395 C99-932283-4

00 01 02 03 04 TC 6 5 4 3 2 1

Printed and bound in Canada
Set in Monotype Plantin Light

To the memory of all the people in Sydney who lost their lives
through the making of steel—the workers, their families
and all those exposed to the toxic legacy.

Contents

Acknowledgements

This book has been a labour of love. Despite all the anger we feel for the decades of neglect and betrayal we have documented, our overwhelming feeling is one of love and solidarity with the people of Sydney, even those who doubtless will find something here with which they don't agree.

Many people were of invaluable assistance, sharing with us the stories of their lives. We particularly thank Clotilda and Dan Yakimchuk (who also let us stay with them, turning their home into our office), Juanita and Rickie McKenzie, Debbie Ouellette, Ronnie and Debbie McDonald, Louise Desveaux, Barbara Lewis, Eric and Peggy Brophy, Clyde Hoban, Don Puddicomb, Pat Wall, Nelson Muise, Gordon Kiley, Lorne McIntyre, Donnie MacPherson, Ed Johnson, Donnie Gauthier, Ada Hearn, Larry Nixon, Anne Ross, Doug Clyke, John and Cindy Steele, Ron Deleskie, Don Deleskie, Mike Britten, Kaz Siepierski, Mike Ferris and Marlene Kane.

Thanks to songwriters and poets Max MacDonald, Leon Dubinsky, Shirley Christmas (who also shared her life story) and Kenzie MacNeil (who also assisted enormously in tracking down song fragments and editing text). Editing help from Glen Davis, Howard Epstein, and Geoffrey and John May is also much appreciated.

We are grateful to photographer Warren Gordon for all his work on our photo section.

Linda Pannozzo, Dan Bunbury and Paul MacNeil helped with our research, tracking down obscure reports and background information. Elizabeth Beaton and Kate Currie of the Beaton Institute at University College of Cape Breton provided valuable and ever-cheerful assistance. Ron Crawley of the Canadian Union of Public Employees shared his personal story and professional expertise and advice.

Angela Rickman and Christopher O'Donnell at the Sierra Club of Canada Ottawa office helped a great deal with researching Chapter Ten, and Patricia Armstrong of the Council of Canadians coordinated the various stages of the manuscript with her usual grace. We are grateful to Farley and Claire Mowat for inspiring us to write the book in the first place and to Phyllis Bruce, our editor, for her expert guidance, patience and support.

Throughout the book, there are moments when the Sierra Club of Canada played an important role. To avoid bogging down the writing, we opted mostly to omit references to Sierra Club. However, it should be noted that the Sierra Club of Canada has devoted considerable resources and time to the struggle to relocate residents and remediate the Muggah Creek watershed. Many of the events recounted in the book involved the Sierra Club, from organizing against the "bury it in slag" plan, to pressing Sergio Marchi to visit the site, to initiating a major project involving the twinning of a Superfund community in Georgia with Sydney, to producing a video documentary, and commissioning the International Institute of Concern for Public Health to conduct a peer review of the CANTOX report on Frederick Street. Sierra Club of Canada is the only national environmental group to have championed the cause of cleaning up the tar ponds and we are grateful for the group's ongoing commitment.

We both wish to thank our families for their ever-constant support, particularly Maude's husband, Andrew, and Elizabeth's daughter, Victoria Cate, who so willingly and cheerfully shared her mother with this labour of love.

MB and EM

Our dwindling forest lies in silence
emptied of life once known,
As the winds rage in anger for
its creatures are all but gone.
Only its haunting cries are
heard today.

And the tears do flow . . .

The great eagle flies endlessly
searching for its home,
While the rivers and streams
pass by filled with grief,
For it no longer carries
life once abundant.
Only its haunting cries are
heard from far away.

And the tears do flow . . .

The drums of our Nation
are no longer heard.
Our songs of ancient
times are forgotten.
Only their haunting cries
could be heard today.

And the tears do flow.

—Shirley Christmas Kiju Kawi,
Within My Dreams

The Sydney Tar Ponds and Surroundings

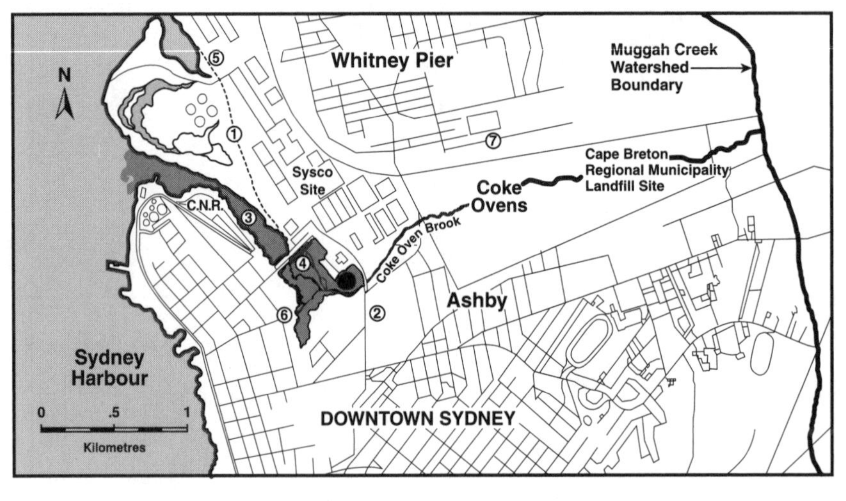

1 Original water line
2 Don Bosco School
3 North Tar Pond
4 South Tar Pond
5 Tar Ponds Incinerator
6 Intercolonial Street
7 Frederick Street

We have the arsenic, we have the naphthalene, we have the lead. The ground is poison, the air turns your lungs raw, now there is orange goo oozing across the cellar. Welcome to Sydney.

—Debbie Ouellette, Frederick Street resident

Introduction

Spring 1998 was unusually warm. Juanita McKenzie had driven home from work one balmy evening and remembers how pleasant it was after a harsh winter to shed her heavy winter wools for a light spring coat. She pulled up in front of her Frederick Street home in Sydney, Nova Scotia, and stepped out of the car. Across the street, she saw a scene out of a grade B science fiction movie. Behind her neighbour Debbie Ouellette's house, two men, dressed in sealed white E.T.-type environmental hazard suits, complete with breathing apparatus, were posting a sign that read Human Health Hazard. Juanita McKenzie looked down at her cotton shirt and pants and asked herself, "Am I underdressed?"

For months, Juanita, Debbie and their families had been sick with every kind of ailment they could imagine and some they couldn't. Kidney infections, nosebleeds, nausea, diarrhea, headaches, tingling joints, ear infections, bloody stools, bloody urine and severe coughs were sweeping through the families that lived in the 17 homes of Frederick Street. All the dogs had died, one after it had literally glowed in the dark. Deformed mice, with batlike heads and kangaroo-like feet, had appeared. Lilacs and roses had bloomed pitch black and then disintegrated. One day in early May, when Debbie Ouellette was cleaning up her backyard, she noticed a bright yellow goo seeping out of the embankment directly behind her house and fluorescent orange chemicals lighting up the creek that runs through her property. Juanita and Debbie were terrified.

Perhaps, they now think, they shouldn't have been surprised. Residents of Frederick and nearby Tupper streets, and Lingan Road as well, had been dying of cancer in disproportionate numbers for years. Cancer was their uninvited, dreaded and constant companion.

For their homes border the worst toxic site in Canada and arguably the worst in North America. Behind Debbie Ouellette's house is a 3-metre-high chain-link fence surrounding the 50-hectare heavily contaminated coke ovens site that drains to the infamous Sydney tar ponds, the toxic legacy of 100 years of steelmaking. The coke ovens site, polluted to depths of 24 metres, contains uncalculated amounts of deadly PAHs (polycyclic aromatic hydrocarbons, the largest group of cancer-causing chemicals in the world) and heavy toxic metals. The estuary contains 700,000 tons of toxic sludge, a witch's brew of carcinogenic chemicals 35 times worse than New York's infamous Love Canal, Hooker Chemical's abandoned toxic site upon which a housing subdivision was built. For years, the residents of Frederick Street lived with their fears and the unconfirmed risks to their health.

But suddenly everything seemed immeasurably worse. When Environment Canada officials confirmed that the backyard soil and brook running behind the homes contained arsenic and other deadly chemicals in concentrations many times over the allowable limit, the women knew that they and their children were in mortal danger. "My heart hit the floor," says Juanita. They also knew they had to take action. This is their, and their community's, story.

We are an island, a rock in the stream,
We are a people as proud as there's been.
In soft summer breeze or in wild winter wind,
The home of our hearts, Cape Breton.
—Kenzie MacNeil, "The Island"

CHAPTER ONE
Paradise Lost

He craned his long neck forward. Around his elongated insectlike legs, cool water lapped, the sun glistened and fish swam unawares. His focus was disciplined. Nothing would intrude on his all-encompassing mission. In a single flash of uncoiled reflex, his beak sought out the prey. Patience rewarded, the great blue heron finished the meal, extended his wings and flew to the sun.

This was all decades ago, in the marsh grasses and rushes of a lovely quiet place, in an estuary with a broad and deep channel. The estuary gathered the waters from a hillside on an island and carried them out to the Atlantic Ocean. The island itself reared out of the sea like a breaching whale—all curves and turns and blue-green flashes of creation.

Cape Breton Island is undoubtedly one of the most breathtakingly beautiful places on earth. It protrudes like a lobster's claw from the eastern tip of Nova Scotia's mainland. A causeway links the island to the mainland, but the political and cultural links are more tenuous.

The island, roughly one-third the size of Vancouver Island, not only is surrounded by ocean, but has large watery expanses where—to use the lobster image—the pincer is separated from the claw. Cape Bretoners boast that the Bras d'Or Lakes form the world's largest saltwater lake, but they could as easily proclaim the Bras d'Or as the world's smallest sea. Although Bras d'Or means "arm of gold" in French, the name actually came from "Labrador," the term that Portuguese explorers used to describe much of Cape Breton.

Northwest of the lakes are the Cape Breton Highlands, still a wilderness within the national park but heavily clear-cut just beyond park boundaries—most of the area is ringed by the legendary Cabot Trail. Enormous whales, majestic bald eagles and comical, colourful puffins all find scarce habitat there. Forests and pastures and mountains rise spectacularly from the sea, levelling to plateaus of forest and bog. Many people say the island is "God's country."

Not surprisingly, people were attracted over the centuries to the bounty of Cape Breton. Forests were felled, fish harvested and outpost trading posts established. The estuary where the blue heron fished long ago was found far from the highlands, just at the tip of the lobster's pincer. Mussels, clams and fish of all kinds flourished in the nutrient-enriched waters of the half-sea, half-creek ecosystem. The wealth of those waters appealed to the indigenous people of the area, who fished and hunted along the estuary's banks and in its waters. After the early wars between English and French forces for domination of the island ended with the fall of the walled city of Louisbourg in 1758, English inhabitants put down roots. The natural harbours of the northeast side attracted the founders of Sydney more than 200 years ago.

A contemporary of Alexis de Tocqueville, M. Arthur de Gobineau, visited Sydney in 1859. His journal is full of superlatives:

It is an immense sea-shore which rises like an amphitheatre by prolonged undulations up to moderate heights. On the horizon you see long, harmonious

> lines uniting mountains to hills which stand out notably against the sky. . . . On the shore, on my right, could be seen nice country cottages, gay and pleasant, surrounded by fences. . . . on the left, a series of tree shaded houses led to the town proper, built of wood, as clean and charming as Saint Pierre [a French protectorate island south of Newfoundland] is sordid. . . .
>
> Under such a magnificent sky, such a beautiful sea, without any regrets, we were, for the moment, perfectly happy and even more so because Sydney is considered to be the paradise of this part of the world.[1]

The town grew, benefiting from its suitable harbour, its fishery and its trade in lumber. The Mi'kmaq, the aboriginal occupants of a vast territory that included parts of what are now Maine, New Brunswick and Nova Scotia, had long relied on the wealth of the sea and its shores. Although known as a warlike people, they had evolved as a society rich in story, legend and religion. The people of Glooscap, as their creator is known, had been willing traders as the strange new settlers arrived. They accommodated the new people peaceably, and the French soon determined that the Mi'kmaq word for cow was "French moose."

The usual pattern of European subjugation and expropriation of the best lands followed, and the Mi'kmaq people were soon restricted to reserves. Most communities remained in traditional lands along the Bras d'Or Lakes, but one Mi'kmaq community settled within Sydney, along what became known as Kings Road, the main land route leading to the city. It was an ideal spot, as it provided water access to the arm of the harbour and offered a good place for a trading post, where the Mi'kmaq could sell their pickaxe handles, handwoven baskets and other goods to passing travellers along the road. In 1877, the federal government set the boundaries of the reserve—a grand total of two acres (less than one hectare).

Every summer, the Mi'kmaq moved to their customary summer fishing camp along the banks of a rich estuary in the southwest corner of the growing city. As settlers arrived, the Mi'kmaq profited from trade with the European sailing ships, and then the Bristol steamers carrying tins and metals to be exchanged for eels and baskets. Scottish settlers had largely converted the banks of the estuary to farms. The creek that ran through one large holding took the name of the farmer, Henry Muggah, and became Muggah Creek (its Mi'kmaq name was Cibou, meaning water). The Scottish name eventually extended to the whole of the watershed and the estuary itself.

Farms sprang up around Sydney, with sheep dotting the hillsides. The early settlers faced short growing seasons and poorer soils than in much of Canada. Life was hard, and most families relied on a milk cow, chickens and a subsistence garden to survive. Men typically supplemented the family income by stewarding a woodlot.

But the pastoral beauty of Sydney and its surrounding communities was soon to end with a dramatic shift in the economy toward the end of the nineteenth century. The riches of the forest and sea were to be eclipsed by the black gold of the deep underground.

The National Dream

In the late nineteenth century, Cape Breton joined the Industrial Revolution with a vengeance. Steam was replacing wind. Coal was the new driver of economic growth, and Cape Breton had a lot of it. Geologists identified its rich reserves as the Pictou coal formation, extending from the mainland of Nova Scotia to pockets on Cape Breton at Inverness and in areas near Glace Bay, North Sydney, Sydney Mines and Dominion. Coal shafts were sunk, and men and boys went to work digging at the rock surface. Pit ponies struggled to bring the fossilized energy of the sun to the surface.

With the return to power of Canada's first prime minister, Sir John A. Macdonald, in 1878, an era of economic nationalism was launched. Macdonald's National Policy—to build Canada from east to west, improving economic autonomy—is now remembered more as the "national dream" of a railway. By 1891, rail lines linked Sydney to Halifax, making it possible for Cape Breton coal to move to industrial centres elsewhere. Within a year, production had been boosted to 1.5 million tons of coal.

The demand for coal seemed insatiable. Captains of industry decided that the small collection of coal mining companies in Cape Breton needed to be consolidated to achieve higher levels of production. In 1893, the first major industry player emerged, as all coal mines joined to form the new Dominion Coal Company (DOMCO). Ironically, it was to be headed not by a Cape Bretoner, nor even a Canadian. The pattern for sawmill and logging operations in the Maritimes was the same; in fact, local resources were generally exploited by people "from away." Despite Sir John A.'s best efforts, the Maritimes' strongest link in trade and development was not to Ontario, but to the "Boston States." Thus, the name imprinted on generations of Cape Bretoners was that of a powerful, aggressive and above all ambitious Bostonian—Henry Melville Whitney.

As director of the Dominion Coal Company, Whitney soon achieved control of all the coal seams east of Sydney Harbour. By 1913, production levels had soared to over 6 million tons a year.[2] But this was only the beginning. The industrialization of Nova Scotia had led to early attempts at manufacturing iron and steel, and mainland Nova Scotia had two such operations, one begun in New Glasgow in 1872 and one started in Trenton in 1883. Turning his eye to the possibility of capitalizing on his coal empire to branch out into steel, Whitney sought out a partnership with a prominent steel manufacturer from Ohio, Arthur J. Moxham. The two men raised private capital of $25 million and created a new corporation, the Dominion Iron and Steel Company, competing with the New Glasgow and Trenton operations. Major shareholders were the president of the Canadian Pacific Railway, Sir William Van Horne, the president of the Canadian Bank of Commerce, George Cox, and Montreal financier James Ross.

From its earliest days, steelmaking in Cape Breton relied on government largesse. Whitney and Moxham were able to secure grants from the federal, provincial and municipal governments. As well, the federal government provided them, free of charge, with 500 acres (approximately 200 hectares) along Sydney's harbourfront, as well as a 30-year holiday from provincial and municipal taxation.[3]

The enormous quantities of water required—65 million gallons a day, or approximately 300 million litres, were provided free of charge, and the cost of coal was far below the norm of 12.5 cents a ton at only 6.5 cents. The federal government bought the Muggah farm and other properties in that 500 acres and gave them to the Dominion Iron and Steel Company, or DISCO as it became known. Needless to say, the Mi'kmaq lost their summer fishing camp, without compensation. The government also expropriated a third of their land—with compensation—to make room for the railroad, restricting the community to 1.4 acres along Kings Road.

The huge grant of land to Whitney's mill changed the town of Sydney forever. As one commentator has noted, the grant represented "some of the choicest lands within the town's boundaries, and [cut] the town into two isolated halves. To this day only a single street, which passes over the steel plant by a long overpass, connects these two halves."[4]

On one side of the divide, Whitney's name attached itself to a community that grew as workers were attracted to jobs in the new industry. Whitney Pier was cut off from the rest of Sydney by divides of class, culture and politics. But even on the other side of the overpass, the closest community, Ashby, was built right up against the edge of the coke ovens. Sydney was now a steel town and Henry Melville Whitney, its most influential businessman.

A Steel Plant Is Born

On July 1, 1899—Dominion Day—the first sod was turned with ceremonial flourish for the new steel plant along Muggah Creek. Like other corporate kings of the era, Whitney and Moxham made grandiose plans. Whitney was cocksure of success. "I believe," he said, "that the establishment of these iron works will be the means of introducing Sydney to the length and breadth of the whole world. I cannot control my enthusiasm when I think of the future."[5]

They set in motion the construction of four hundred coke ovens, four 250-ton capacity blast furnaces and ten open-hearth furnaces.[6] They secured abundant raw material for making steel. Whitney already had the coal required, but they needed a supply of raw ore. Some was available locally, but additional supplies would have to come from the iron ore and limestone mines of Wabana, Newfoundland. Whitney bought the mines so that he would control all the elements of production.

If Cape Breton is God's country, the building and functioning of coke ovens, blast furnaces and open-hearth furnaces brought a little bit of Lucifer's business into that part of creation. For how else can we visualize the molten iron and steel, the melting coal, the temperatures reaching thousands of degrees? Regardless of their denomination, and despite their fond memories of a camaraderie and solidarity, the men who worked in the steel mill and coke ovens recall it as hell on earth.

Steelmaking is a complex process, relying on the right mix of numerous factors. The recipe must be perfectly calculated and prepared through a complicated set of stages, using both combustion and chemical processes to turn iron ore, limestone and coal into steel. The first step is making coke, a form of carbon that is both hard and porous.[7] Coke creates an intense, smokeless heat as it burns. To make coke, coal is fired in ovens without the presence of oxygen, at 400° Celsius. DISCO's coke ovens were designed to capture the by-products as well, because when the coal is reduced to coke, numerous gases and tars, with other industrial uses, are released.

The coke ovens were rectangular and built in long rows, called batteries. Between each oven was a series of heating flues designed to keep the walls hot. If the walls ever cooled down, the thick brick would crack apart. When the walls hit temperatures of 1,100° Celsius, the ovens were ready to be loaded through doors at each end. As the grey clouds of coking gas rose, men on the roof, wearing thick wooden clogs over their boots to keep them from catching fire, sealed holes in the roof with mud to keep the ovens airtight. The doors were then sealed for the full coking cycle of about 17 hours.

Once the coke was ready, the doors at either end of the oven opened and a giant steel ram scraped through the rectangular oven, pushing the molten mass of coke to the railcar below, known as the "quenching car." As the coke hit the open air, it burst into flames and flowed like flaming lava into the railcar, which transported it by rail to below the water tanks, to be quenched with gallons of water. Immediately, huge clouds of toxic gas would billow out, spreading a chemical mix of tars, oils, naphthalene, ammonia, phenols, cyanide, sulphide, thiocyanate and polycyclic aromatic hydrocarbons (PAHs) to the workers and the neighbouring community. The water then carried tarry poisons to the nearby brook.

Once quenched, the coke was ready for the next steelmaking stage. Operations moved to the blast furnace, which had two main purposes: to reduce iron ore to a metallic iron called pig iron and to remove impurities. The blast furnace was a cylindrical chamber about 8.5 metres wide and between 24 and 30 metres high, surrounded by walls of fire brick over a metre thick. At the base of each hearth were tap holes to collect the molten iron and waste slag material.

The main ingredients in this recipe were iron ore, coke and limestone, all melted at incredible temperatures through blasts of forced super-heated air. The melting zone of the packed cylindrical chamber moved from near the top, where the hot air was blasted, through to the bottom. As the iron ore heated, it became a porous mass, melting through the coke that held it in place, until it reached temperatures that forced it from solid to liquid and melted down to the base of the furnace. Lighter than molten iron, the waste materials, impurities and lime from the limestone floated to the top. This mixture of substances, called slag, was tapped from one hole, where it ran out to a deep pit. Later, it was dumped into the estuary in slag piles, from which materials for road fill or aggregate for concrete could be "mined."

The molten iron—now called pig iron—was tapped lower down, in a process called casting. It flowed like a fiery snake along a trough in the sand to a train of ladles that shifted the melted iron to the next set of ovens—the open hearth.

To make steel, the pig iron had to be further refined to reduce the amount of carbon. The open hearth had a tilted saucer-shaped floor where the iron was "open" to the flames below. Inside the open hearth, at temperatures of 1,600° Celsius, the molten pig iron became molten steel.

The steel was tapped from spouts at the lowest part of the hearth and transported to a series of further processes. Although it could be shipped out as steel ingots after the open hearth stage, more often it went to the blooming mill, where ingots were rolled into "blooms," or hot rods of steel for future refinement. There, the steel was reheated for 8 to 12 hours to achieve the uniform temperature that prepared it for the rolling mill. At the right temperature, between 1,100° and 1,300° Celsius, it was soft enough to be squeezed through a wringer into different elongated shapes.

As the steel reduced in thickness, it extended in length. Wild pieces of hot steel could fly out of the wringer and kill a man faster than his friends could warn him. But somehow generations of workers learned the tricks of "catching" steel. This was the dangerous process of carrying steel billets from one set of rolls in the rod mill—essentially, the men had to catch the white-hot rods and swing them around their bodies to the next stage of production. Over the years, other mills were added to refine the steel further into bars, wires and nails, billets and rods.

The Compromises Begin

All of these steelmaking processes required high-quality ingredients. Too much sulphur in the coal would make an inferior calibre of coke. Impurities in the iron ore would require far more limestone to remove them as slag. Incredibly, however, neither Arthur Moxham, now vice-president and general manager of DISCO, nor Henry Whitney stopped to test the quality of their ingredients. In a rush to begin full operation, they failed to run the most basic tests on their coals and ores. DISCO tapped its first steel on New Year's Eve 1901 and right from the start there were problems.

The coal from Cape Breton seams was very high in sulphur, so far more coal had to be baked to produce usable coke. The iron ore from Wabana was full of impurities, such as silica and rock, so far more limestone was required to pull out the impurities as slag. The unusually large amounts of limestone required in the blast furnaces caused the furnace linings to deteriorate rapidly. The problem was "severe enough to cause the lining of No. 1 furnace to burn through after a few weeks of operation."[8]

The poor quality of the basic ingredients led to higher costs, less marketable and inferior products, and far more waste. In what had been Muggah Creek, the slag would eventually create a mountain range of waste, stretching hundreds of feet high and reducing the mouth of the estuary by nearly a mile. With more coal being baked to make less coke, the mills were producing larger quantities of tarry sludge, benzenes (volatile liquid used as a solvent), ammonium sulphates, naphthalene (a white crystalline substance used to make mothballs) and PAHs. These by-products and wastes were simply dumped into Muggah Creek, or into what became known as Coke Ovens Brook, which runs to Muggah Creek.

Not only were the raw materials of poor quality, but the initial management of DISCO was a disaster. The construction of the plant was plagued by waste and error. It was estimated that the plant could have been built for less than two-thirds what Moxham and Whitney spent, with between $7 million and $8 million wasted.[9]

In an attempt to reduce "impurity levels to a uniform low level,"[10] DISCO began experimenting with retrofitted blast furnaces lined with an acidic material. These were designed to avoid the use of limestone and to melt the ore through multiple heatings. But the results were not good. The pig iron was of very poor quality and the steel emerged pitted and inferior. A metallurgical journal analysis of these early days reported "excessive erosion of furnace linings, high scrap losses, low ingot yields, uncontrolled silica content, and steel suitable for soft steel use only."[11]

Meanwhile, another steel plant—the Nova Scotia Steel and Coal Company (SCOTIA)—had started in Sydney Mines. Owned and run by Nova Scotians, SCOTIA had a well-integrated operation. Unlike Whitney's "super-mill," SCOTIA made only pig iron and steel billets, which were then shipped to one of the original Nova Scotia mills, now amalgamated in the same company based in Trenton, Nova Scotia. At its peak, SCOTIA employed 6,500 workers and operated a huge hydraulic forging press making the world's largest steel shafts. Yet government attention seemed to focus more on the foreign-owned DISCO.

Early efforts by Whitney to merge his coal company with SCOTIA's steelmaking were rebuffed by a management known for its prudent approach to business. Whitney was a man of action and, with Moxham, he believed that anything was possible. Initial failure did not deter these champions of capitalism. They turned to the federal government and demanded more money. As recounted in an excellent summary of the steel industry's toxic legacy by John McMullen and Stephen Smith, "In a refrain that was to become familiar, [Whitney] warned of regional disaster if the plant closed. A state 'bailout' was arranged and the Dominion Iron and Steel Company continued to produce low grade, toxic steel."[12]

By 1903, Whitney had made so many costly errors that he was forced to sell his shares in the Dominion Coal Company and the Dominion Iron and Steel Company to Montreal tycoon James Ross and his Canadian Pacific Railway (CPR) companies.[13] While no one in Cape Breton knew it, the coal mines and the steel plant were now essentially owned by the same corporate interests, the Bank of Commerce and the Bank of Montreal.[14] The poor quality of the coal soon led to a falling out between the two companies now controlled by Ross and his associates. Shareholders in DISCO wanted to demand higher quality coal from the Dominion Coal

Company operations. The whole mess landed in the courts, eventually reaching the highest court—Britain's Privy Council. The Privy Council ruled in favour of DISCO and ordered a huge $3.5 million settlement from DOMCO to DISCO and the effective takeover of the coal company by the steel company.

In 1910, the two companies were amalgamated into a single new holding company, Dominion Steel Corporation, with J.H. Plummer of Toronto installed as president. Within several years, the First World War created strong markets for Sydney's steel and the company finally became profitable. A coking by-product called toluene could now be used to make explosives, including TNT. An unprecedented era of profitability led to further mergers and the formation of the British Empire Steel Corporation (BESCO) in 1920.

BESCO

At that time, the creation of BESCO was the largest corporate merger in Canadian history. It amalgamated the holdings of Dominion Steel Corporation, including its coal operations, Nova Scotia Steel (ending the life of SCOTIA, the only Nova Scotian–held steel operation), the Halifax shipyards, the Wabana iron mines of Newfoundland and the Eastern Car Company. Its new kingpins were Colonel Grant Morden, a director of the Canada Steamship Company, and an American wheeler-dealer named Roy Wolvin who had assumed control of Dominion Steel just prior to the merger.

For all its corporate bravado, the merger was in trouble from its inception. The corporation had been excessively over-capitalized, offering more shares than its assets were actually worth. No new money was invested in a transaction that was essentially an exchange of shares. The merger drained the Cape Breton steel operations of all their wartime profits while shareholders clamoured for dividends that BESCO could not afford. No repairs or modernization were possible. During the Depression years, the plant slowed operations to nearly a standstill, increasing the strain on production and on the lives of the workers. Working conditions deteriorated as the plant became increasingly dirty and dangerous. By 1924, the ongoing financial crisis was so severe that BESCO could not even cover its overhead.

Wolvin decided to cut loose the failing steel plant by placing the DISCO operations in receivership with National Trust. In 1927, National Trust decided that BESCO and DISCO should be dissolved. But the Nova Scotia courts refused to allow BESCO's dissolution. In the continuing drama of Cape Breton steel, Wolvin sold his interest to a new group of financiers led by Herbert Holt and a group aligned with the Royal Bank.

The acronyms were shuffled again, and in March 1928 a new holding company was set up—DOSCO (Dominion Steel and Coal Company)—with a Royal Bank director, C.B. McNaught, at the helm. Ongoing British interest was maintained through directorships for Sir Newton Moore and Lord Inverain. Further federal government subsidies buoyed the troubled operations. The Canadian Coal Equity Act of 1930 gave industries a bounty of 49 cents a ton for the use of Canadian coal in making iron and steel. With ready supplies of poor quality Cape Breton coal, the legislation guaranteed that DOSCO would keep using it, delaying the switch to higher quality, less toxic sources.

The Second World War raised the flagging fortunes of the steel industry globally, and Cape Breton was no exception. The federal government infused the steel operations in Sydney with a special wartime assistance package of nearly $6 million for a new plate mill, new open hearths and new blast furnaces. But unlike the steel mills of Ontario which had also expanded in the war years, DOSCO experienced only 33 per cent growth after 1944, whereas operations at Stelco in Hamilton tripled and Sault Ste. Marie's Algoma mill expanded fivefold.[15]

It was almost impossible for DOSCO to stay competitive with Ontario mills. Both Stelco and Algoma could sell 85 per cent of their steel within an 800-kilometre radius of their mills. Sydney could count on sales of only 10 per cent of its output within such a radius. Exports, with attendant higher costs, were critical.

Although DOSCO added new coke oven batteries and by-product facilities from time to time, overall the steel operations in Sydney were in constant need of modernization and were still plagued with quality problems. To aid the mill, the federal government responded with the Maritime Freight Rates Act, which reduced shipment costs for rail within the Maritimes by 20 per cent, and the cost of shipping rail beyond by 30 per cent. But this support was to no avail.

Industrial Fall-out

In 1957, DOSCO became a subsidiary of the British multinational Hawker-Siddeley. The next ten years of operations—the last ten years of private sector control—have been characterized as the "virtual dismemberment" of the Sydney steel operations.[16] Any profits were diverted to other branches of Hawker-Siddeley's operations. In 1961, a slump in the world steel market drove the workforce down below 3,000 from the wartime high of 5,400 employees.

Initially, however, the new management was hopeful. With the subsidized rail rates and Hawker-Siddeley's multinational links, new ores could now be obtained to make better steel. Ores from Brazil, Labrador, Africa and the United States were shipped in to Sydney. While the quality of steel improved, the new ores caused local air pollution to get even worse. The Brazilian and Labrador ores, in particular, produced different chemical reactions, creating more oxygen in the blast furnaces. Blast furnace "slips"—rapid increases in gas pressure forcing clouds of reddish pollution out the top of the furnaces—had previously occurred several times a month. Now, with imported ores, the number of slips increased to 117 emissions per month, depositing up to 150 tons of particulates and dust per square mile per month.[17]

Great billowing clouds of reddish smoke shrouded Sydney in a nearly permanent orange-red haze. Children had neighbourhood permission to run to the nearest house, without knocking, to escape the orange dust. The air was loaded with iron oxide, ash and coke particles. At the beginning of a blast furnace slip, the forced gas release would send the red clouds 150 to 450 metres in the air in seconds. A report by the federal department of Health and Welfare noted that the increased slips were causing "deposition and discoloration of buildings, walls, textiles and laundry and any exposed surface." They were also leading to the "penetration of pollutants into the respiratory tract, nasal passages and lungs" of local residents.[18]

These emissions could have been entirely avoided by pre-treating the imported ores in a sintering plant. In the sintering process, the ore is combined under heat with other materials to enlarge its particles to a more granular size. The cost of such a plant was $6 million—a small amount to such a large corporation. But as the only advantages of the expenditure would be improved air quality and better protected public health, DOSCO decided against it, and the government did not insist. The strongest language used by provincial health officials was that DOSCO should "attempt" to reduce air pollution and "perhaps" invest in the new sintering plant.[19]

DOSCO, under Hawker-Siddeley management, also attempted to use higher quality coal, purchasing larger quantities of low sulphur Kentucky coal and mixing it with smaller amounts of Cape Breton coal. This was in concert with a recent Nova Scotia government report setting out options for the future of the Cape Breton steel industry in the face of continued and growing threats from overseas competitors.

In addition to recommending reduced freight rates and "importation of low-volatile coals,"[20] the government report had favoured the option of sinking another $30 million into modernizing the mill—but not to reduce air pollution. DOSCO accepted the government recommendations, with a disastrous impact on the Cape Breton coal industry. Five Cape Breton coal mines were forced into closure in 1960, and with DOSCO now using Cape Breton coal for only 20 per cent of its coking, the local coal industry was in crisis. In 1965, the federal government stepped forward. This time, it assumed direct state-controlled ownership and management of the coal mines with the creation of the Cape Breton Development Corporation (DEVCO).

Another corporate casualty of the switch to less toxic coal was the largest of the by-product facilities on the coke ovens site. During the 60 years of steelmaking, numerous efforts at utilization of coking waste had come and gone—a Scottish fertilizer plant in 1911 (gone by the 1930s), a cement company in 1910. But the largest and most successful of the by-products operations was Dominion Tar and Chemical, now known as Domtar. Domtar established operations in 1903 to capitalize on the abundant coking waste.

The coke oven batteries, designed to allow by-product manufacture, channelled gases to a series of underground pipes. Literally hundreds of kilometres of piping lay below the 50 hectares of the coke ovens site. The Domtar plant produced creosote (a dark brown oil distilled from coal tar), pitch, ammonia and sulphate. Annual production levels for tar were between 6 and 8 million gallons. But with the switch to coal that made better coke, the quantity of available by-product fell. By 1962, Domtar had closed its operations in Sydney.

Despite tens of millions of dollars in subsidies, reduced rail rates and state guarantees, Hawker-Siddeley decided to pull the plug on its DOSCO plant in Sydney. On October 13, 1967, a day that became known as Black Friday, it announced the plant would close in April 1968.

For 66 years the plant had polluted the 500 acres it had been given. It had indiscriminantly dumped huge amounts of toxic waste on the entire community and any area downwind. Owners and acronyms had come and gone—from DISCO and DOMCO to BESCO and DOSCO—as eager entrepreneurs sought to make a killing where others had failed. But the pollution remained.

New names were attached to the surroundings. Sydney and its neighbouring towns of Glace Bay, New Waterford, Sydney Mines and North Sydney were collectively known as industrial Cape Breton. Muggah Creek became known only as the tar ponds, once the black ooze that befouled the coke ovens site migrated to the estuary and made it a dead zone. The mouth of the once wide and pleasant creek leading to the harbour was now shallow, choked by railyards on one bank and the steel mill on the other. Though the mill had been built on the banks of Muggah Creek, decades of slag dumping had filled in so much of the estuary that the plant was now nearly a mile from the shore, with mountains of slag, known as the high dump, lying in between. The daily tidal flows brought the coking tars and other poisons to the open harbour beyond.

In 1968, the steel plant should have closed. If each worker had received a million dollars in compensation, the cost would have been less than the subsidies that poured in to keep the mill afloat. More important, many lives would have been spared. But this was a steel mill that would not die, even if it killed everything around it.

Oh the sweat on the back is no joy to behold,
In the heat of a steel plant, or mining the coal.
And the foreign-owned companies forced us to fight,
For our survival and for our rights.

—Kenzie MacNeil, "The Island"

CHAPTER TWO
Sons of Steel

The men who worked in the Sydney steel plant and the adjoining coke ovens had a love-hate relationship with the corporation of many names. From the day it began operations, the plant had a well-deserved reputation as a ruthless employer with one of the worst safety records in Canada. But the giant company, able to attract huge amounts of foreign and domestic capital, soon established itself as the dominant industry in town and, displacing small independent producers and businesses, turned most of Sydney into a community of wage labourers with few options.

Although the workers would eventually fight for safer conditions in the plant, they and their families learned to work, live and love in what was to become one of the most poisoned environments in the world. The plant became the lifeblood of the community. The billowing orange smoke signalled that the blast furnaces were operating at full tilt, and that meant the men were working. And everyone knew how terrible life could be when the men were not working.

The attitude could best be summed up in the words of one company president who addressed a group of protesting women. Monday was wash day, and the women were unhappy that the company's smokestacks, which coated the clothes on their lines with red grime, couldn't be shut down for a few hours. "No smoke, no baloney!" he thundered and the women returned to their homes.

The Come-From-Aways

When the steel plant opened at the turn of the century, the promise of work acted as a magnet, attracting thousands of job seekers. The population of Sydney grew by 600 per cent in the 20 years after the company's first sod was turned.[1] Single young men came from other parts of Canada, the British Isles and the United States. The company made them welcome and gave them the skilled jobs closed to the immigrants who began flooding in from other "less desirable" parts of the world. The foundations for deep class and race divisions were laid in the Sydney community, particularly in the areas adjacent to the company operations, alongside the first bricks of the steel plant.

Until this time, most Cape Bretoners other than the original inhabitants, the Mi'kmaq peoples, were of Scottish descent, and in many communities only the old Gaelic was spoken. In his autobiography, union organizer George MacEachern, who was born in Sydney in 1904, tells a story about how insular life was before the turn of the century. His mother married a policeman who "drank on occasion" and was, therefore, the subject of much scandal. His grandmother, who spoke only Gaelic and deeply disapproved of her daughter's choice of husband, was sure he drank because her daughter prayed in English and therefore "talked to God in a language He didn't understand."[2]

Over half of the new immigrants came from Newfoundland (not yet part of Canada). A crisis in the ailing fishery forced many young Newfoundlanders to seek work elsewhere, and

Newfoundland migrant labour came to form an integral element of industrial Cape Breton. Steamship companies helped recruit many of these immigrant workers. Desperate for unskilled labourers, the Canadian government paid bonuses to the ship companies for every immigrant they brought to Canada from overseas. Steel mill agents scoured Newfoundland for illegal immigrants, who were encouraged to jump aboard the ore boats travelling from Belle Island to Sydney. DISCO paid these workers the lowest wages it offered and housed them in filthy shacks in the area known initially as the Coke Ovens. This immigrant district would grow in size and eventually become the community of Whitney Pier.[3]

Even though Newfoundland was part of the British Empire, Nova Scotians did not hold its people in high regard. They saw Newfoundlanders as unskilled foreigners who were bent on taking local jobs and had no long-term interest in the community. In 1901, the *St. John's Herald* reported on the trouble these boys were having: "During the past five weeks hundreds of men from Conception Bay and points north went to Sydney, believing they would secure work at once. They failed, and all these are now at Sydney faring as best they can, which is but miserably. The majority of them sought work in the coal mines, but the Nova Scotians and others are openly hostile to our people and they could not secure employment."

Labour historian Ron Crawley, who grew up in Sydney and worked briefly at the plant as a young man in the early 1970s, observed that Newfoundlanders were forced to work in dangerous conditions that local labourers would usually refuse. A 1904 newspaper reported that if a Newfoundlander was asked to work in a dangerous place, he would go and "in nine times out of ten the Newfoundlander meets death." By 1903, 83 men were reported to have lost their lives, 49 of whom were Newfoundlanders. The *Sydney Record* cheerfully noted that Newfoundlanders had a propensity for getting killed. In reporting the death of a young Newfoundland steelworker, the *Evening Herald* commented that he was "only a codman."

When a young Newfoundlander was killed in the Glace Bay

mine, clergyman Walter R. Smith wrote to the editor of the *Evening Telegram:* "Fred was killed working for the company, and the company was supposed to have an inquest held over him, which there was nothing of the kind. There was no doctor to see him and bury him, and where he was buried it was just the depth of a coffin. That's the kind of grave he was buried in; when the frost goes the birds will snatch him up."[4]

Newfoundlanders weren't the only group to be singled out for poor treatment. Records from 1920 show that most men at the plant were earning a dismal 32 cents an hour, but "coloured" workers were paid only 15 cents an hour. The company brought in Black steelworkers from Alabama, Pennsylvania and Tennessee to train the local workers, and then they were expected to leave when the job was done. Those who stayed also settled in the Coke Ovens area alongside former West Indian slaves who were fleeing economic deprivation at home, and new immigrants from China, Italy, Hungary, eastern Europe and Russia. These workers, who had been falsely promised good jobs and comfortable living conditions by company agents, found that they were expected to buy their jobs and (rare) promotions from the steel plant with a bottle of rum for the hiring clerk. They formed the bottom rung of the social structure in Sydney and were forced to take the hardest, most dangerous jobs in the system. Slovak workers, often considered "Bolsheviks" and revolutionaries, were often treated as indentured slaves, and given some of the dirtiest, lowest-paying jobs on the line.

Local workers were generally unskilled and were therefore in direct competition with these immigrants. Bitter complaints against the "foreigners" surfaced. In a letter to the newspaper, a local unemployed steelworker charged that the blast furnace was being run by "Yankees and Alabama coloured folks," and said his experience was that "one Yankee is as good as two niggers and one Canadian as good as two Yankees." Another letter, written in 1902, declared that "any competent honest foreman knows that the Cape Bretoners, Prince Edward Islanders or Newfoundlanders, if properly treated, and with a little instruction, could be made as capable and immensely more reliable than the Alabama niggers who are here today and gone tomorrow."[5]

Black workers, like Newfoundlanders, were often left off the company's official accidental death list. In 1905, four Black workers were killed during a storm when they came in contact with high-voltage wires that had blown down on plant property. When the official statistics for accidental deaths on the steel plant were released that year, they did not include these men.

As settlement around the steel plant grew, the area became known as Whitney Pier. This three-square-mile triangle-shaped community on the northeast side of Sydney Harbour was virtually surrounded by the plant, an odd configuration caused by the original grant of 500 acres. The Pier was enclosed by the coke ovens on the south and west, the open hearth, mills and shops on the far west, and the blast furnaces and piers on the north. Dirt roads took shape with names like Tupper Street, Lingan Road and Frederick Street (where Debbie Ouellette and Juanita McKenzie would eventually live), the latter bordering directly on the coke ovens. As it became settled, Whitney Pier developed as a community of distinct, even segregated neighbourhoods based on ethnicity: Scots lived in "Scotchtown," Slavs in "Hunkey Town," Italians in "Shackville," and Blacks in the "Coke Ovens." Today, Whitney Pier is still completely separated from Sydney—the only physical link between the two communities is a huge overpass that travels over the railway tracks between the steel plant and the coke ovens.

The men lived here in company shacks, 50 men to a building, and every bed was shared with another worker on a different shift. Conditions were so bad in the Coke Ovens area that typhoid fever was prevalent. Reports of drunken "foreigners" roaming the streets looking for trouble were widespread and the number of arrests jumped. Immigrant workers whose wives came over settled with their families in tenement housing scattered along the railroad track. George MacEachern writes about the relentless, grinding poverty: "Looking back, you wonder why they didn't revolt long ago, except that poverty alone doesn't make for revolutions.

It might make for rebellions at times, but there has got to be more to it than that."

The Fight for a Union

Conditions in the steel plant and the coke ovens were as primitive as can be imagined. A worker had a 13-hour night shift or an 11-hour day shift, six and sometimes seven days a week. Workers' journals noted that they spent their entire youth on the job. Wages were among the lowest in North America. There were few holidays, no vacations, no extra pay for overtime, no security and no company-paid benefits for injury, illness or death.[6] There were no pensions and men worked until they died, whether by an accident at age 25 or old age at 90.

Steelworkers were not issued any safety equipment or protective clothing, and only in the 1950s did the company enact a limited compulsory safety program to provide hard hats, safety glasses and proper boots. As former steelworker Syd Slaven explains, all the elements of danger—such as molten metal, poisonous gas, moving heavy equipment, overhead cranes, conveyor systems, high voltage, and noise pollution—exist together in a steel plant.[7]

Officially, the company lost over 300 men in accidents between 1901 and 1993, the greatest number in the first half of the century. Unofficially, as any worker will attest, the number is much higher. In the early years, only deaths that occurred at the site were reported. Injuries that led to a home or hospital death were not recognized, nor were cancer deaths caused by the daily proximity to carcinogens. And, of course, many an immigrant labourer went to his grave unrecorded.

Almost from the day the company opened, the workers started fighting back. Not surprisingly, the first strikes were led by immigrant workers who, because they were not allowed to take skilled jobs, were also barred from the newly established Provincial Workmen's Association (PWA), which was dominated by workers of British descent. In 1903, Italian labourers went on strike to seek some job security and better pay. The company claimed that the pickets were "interfering with English-speaking workers going inside the plant," and the PWA did not intervene when police broke up the strikers with "considerable clubbing." The company punished the strikers with beatings, blacklisting and evictions.[8]

The first plant-wide strike occurred in 1904 after the company announced wage reductions, and was led at first by Italians, Hungarians and Newfoundlanders. But this time, the PWA fought alongside the immigrant strikers. PWA grand secretary Moffatt appealed for support from the rank and file at their lodges: "It is the battle of millions of dollars waged by millionaires against men already impoverished. Is money to triumph over men? Is capital to be forever permitted to create conditions unbearable to the great army of toilers?"

The city closed the barrooms on the first day of the 1904 seven-week strike, and the provincial government called in the Argyle Highlanders, armed men of the 63rd Rifles and the 66th Princess Louise Fusiliers from Halifax to break the strikers, who, reported the *Herald*, "were armed with clubs and bludgeons and . . . looked as if they would not hesitate to use the weapons of warfare in their possession." It was a complete rout. Federal deputy minister of labour William Lyon Mackenzie King was brought in to arbitrate the dispute. The company refused to increase wages back to their pre-strike level, but agreed to reinstate all the striking employees in their former jobs and not discriminate against members of the union.

The company immediately broke its word. Leaders of the strike were blacklisted and the union crushed. In its first big confrontation with its workers, DISCO succeeded in demonstrating its total power over the community; its ruthless reaction was seen as a major setback to the working class movement.[9] Coal miners struck in 1909, in 1917 and again in 1925, when they burned the company stores. Each time, they were met with the armed intervention of the state. But in 1917, a local of the Amalgamated Iron, Steel and Tin Workers of America quietly emerged at the steel plant and grew in militancy for the next six years.

The company used other, more persuasive tactics to try to keep out the union. In 1922, it decreed that, from then on, December would be "Safety First Month." George MacEachern remembers a well-dressed company man standing on a table, lecturing them on how much the employer cared about their safety: "My God, the carnage in December. Gee! A friend of mine, who just started around the plant the same time I did, got squeezed to death up at the coke oven. And old Perry, he was chasing a crane when a billet fell on him and squashed him. A great year for squashing . . . . All over the plant there were accidents . . . . Then they put up signs advising you, Don't Get Hurt. Now, this was a contribution, I must say."[10]

In June 1923, the company, now called BESCO, threatened the workers with another wage reduction. This precipitated one of the bloodiest strikes in Cape Breton history. BESCO's official state-backed policy was that "trade unionism is wrong in principle and will not be tolerated." The company hired replacement workers, surrounded the plant with barbed wire and set up military tents. It even mounted a machine gun on the roof of the steel plant aimed at the strikers. Retired steelworker Frank Smith was a young boy of six at the time and remembers the conflict in vivid detail. Strikers would march any replacement workers they could find through the town to the cries of "Scabs! Shame!"

The strike culminated on July 1—later known as Bloody Sunday—when armed military and police forces attacked workers and their families as they walked home from church. One eyewitness recounts the scene: "I saw them coming down Tupper Street four and six abreast . . . and the women taking their children off the street terrified of these army people. . . . On Sunday night those provincial police rode down the people at Whitney Pier . . . most of whom were coming from church. An old woman was beaten. . . a nine year old boy was trampled. . . a woman gave premature birth after she was beaten over the head." Frank Smith also remembers that terrible day: "What I saw, in the roadway before Nathanson's Store, no more than a hundred yards down the road from where I stood, was a crazy blur of horses and riders coming on and a crowd of screaming people running off in all directions."[11]

Once again, the company completely crushed the young union. Some strikers were jailed, many blacklisted and all intimidated. BESCO obtained a legal ban prohibiting groups of strikers greater than three from picketing its premises. The community was deeply divided and the hardship to the families of the workers was severe. Class hatred had grown as well, with the business community siding with the company and the churches paralyzed with inaction. For years, families would remember which side others had taken. Frank Smith recalls the "unrelenting and permanent hate," not only between striker and scab, but between their children as well. In addition, the steelworkers soon had to contend with the Great Depression.

Cape Bretoners are fond of saying that they never noticed the Depression: what happened to the rest of North America in the dirty thirties was daily life on the island. But that's just not true. Bad as living conditions were in the 1920s, they became immeasurably worse in the decade that followed. Massive layoffs at the plant left many families destitute, and with no government help to transport food from the surrounding farms to the people of Whitney Pier, the weekly food vouchers were not always worth much. The $1 voucher received by an unemployed single man could buy him a turnip and a can of milk one week and a little butter and salt cod, but no milk, the next. Families received a $3 weekly food voucher, but had to pay for their own rent, lights, telephone and fuel. Milk for the children was also not provided.

Tuberculosis spread through the community. Men were arrested for stealing coal. Early in the Depression, a small group headed by unemployed steelworker Dan MacKay started an organization for the unemployed. They fought for food and housing, set up a committee to provide milk for children, and waged a militant fight against evictions. Roving squads of the organization's members would sit on furniture to prevent landlords from moving it into the streets.

The Union Is Born

These years hardened and politicized the workers of Sydney Steel. Many of the principal players who fought for the unemployed were instrumental in setting up the Steelworkers' Organizing Committee. George MacEachern demonstrated the new militancy and backbone at the plant during a visit by one of the company's British directors, Sir Newton Moore. The plant council of workers insisted that Moore meet with them, a request that was accepted. Everyone was nervous, bosses and workers alike. When Moore entered the room, MacEachern refused to stand. "You'd think it was the Second Coming instead of the arrival of a g.d. British capitalist," he recalls.

During the meeting, Moore tried to intimidate the men, warning them that the international market for steel was drying up. MacEachern didn't believe it for one second. He wryly notes that "in 1935, the whole world knew, with the exception of one or two of the English aristocracy, I suppose, what Hitler was up to." He was shocked when Sir Newton admitted that the company, now called DOSCO, had sold 500,000 tons of ore to Germany; for years after, the union organizer loved to say in speeches that this sale was the company's contribution to the war against socialism. At the end of the meeting, MacEachern was the only person Sir Newton approached. Moore stuck out his hand and said, "You'll get along." MacEachern looked the British boss square in the eye and said, "I'm not going anywhere."

Local 1064 of the United Steelworkers of America, an affiliate of the Congress of Industrial Organizations (CIO)—labour central in the United States—was born on December 13, 1936, at a meeting in the Temperance Hall in Whitney Pier. Within three months, the union was 3,000 strong. The union fought for and secured an eight-hour day. In 1937, the steelworkers co-operated with the coal miners of Nova Scotia to win the first provincial trade union act in Canada, mandating a "dues check-off," whereby every worker must contribute to the union. They fought and struck for seniority rights.

So organized was Local 1064 that an open hearth union activist of the time describes an elaborate signalling system the men implemented to spread the word of a strike. Key activists in each department would communicate by picking up the departmental telephone and telling the worker in another department, "The bird has flown." This was the cue for workers in the next department to strike if possible.[12] The union also began to influence larger issues and events in Whitney Pier and Sydney. It supported workers at other work sites and called for improved social assistance. Union leaders joined alternative political parties such as the Co-operative Commonwealth Federation and the Communist Party of Canada, an indication of the community's lively and radical political life. Many also ran for municipal office.

In 1940, the company signed its first contract with Local 1064. In retired steelworker Frank Smith's words, it was "hardly a world shaker," and the company immediately looked for ways around its commitments, leading to several important strikes in 1943 and 1946.

Nor was the union free of internal problems. Many resented the American union bosses, particularly CIO leader John L. Lewis (whom some blamed for betraying the workers, in the 1923 strike, by making a deal with the company). An all-Canadian independent union movement was crushed, with a lot of help from U.S. headquarters in Pittsburgh. "Radicals" and members of the Communist Party were marginalized within Local 1064, and purged from the upper echelons of the national section of the union. This left Local 1064 less resistant to management shenanigans in the following decades, both through the boom times of the post-war period (when the plant employed 5,400 workers) and the massive layoffs of the 1950s and 1960s.

Nevertheless, the 1940 contract was a first and significant milestone in the long struggle for justice at Sydney Steel. Men dedicated their entire off-plant time to the union. As Frank Smith notes, "Nothing came easy—All had to be fought for and unity, more unity and again unity was ever key to victory."

Life on the Line

In spite of the (painfully slow) improvements to wage rates, working hours and safety rules at the steel plant over many years, the workers universally recall day-to-day life on the line as dangerous, dirty and exhausting. The company continued to use outdated and hazardous equipment right up until the 1980s. Even with some technological change, the jobs and working conditions stayed remarkably constant throughout the decades. A First World War veteran said that war was less dangerous and more fun than life at the coke ovens, and veterans returning from the Second World War echoed the same sentiment. In fact, many describe their work site as a kind of war zone, one in which, just as in a war, they had to bond with their "buddies" for survival.

Their personal testimonies are filled with stories of courage, danger and humour. Friendship kept them going. Nelson Muise, who worked at the blast furnaces for 42 years until 1983 when he had a heart attack, and whose son, David Muise, is now the mayor of Sydney and a cancer survivor, quips that "you didn't have to be crazy to work there, but it helped." When he describes his years at the company, Nelson's eyes mist over with memory. "We were like a family," he says. "We had the greatest camaraderie you can ever imagine in a group of men. You had to depend on your buddies. Everybody had buddies. The greatest thing in the world was your buddy."[13]

The company was life. "We had steel for breakfast, dinner and supper," says retired steelworker Pat Wall. Ron Crawley Sr. is a prime example of a man who lived his entire life around the rhythym of the plant. His father was a steelworker who at the age of 14 emigrated from Newfoundland during the First World War to work at the plant. He and his wife raised Ron and seven other children in Whitney Pier within sight of the plant. When Ron married and began work at the plant in 1950 at the age of 19, he and his wife, Delores, raised their family in a company house on Victoria Road adjacent to the plant on the Ashby side of the coke ovens where Delores's family had lived for years. Ron and Delores bought their company house for $2,600.

Ron Sr.'s life revolved around his family and the plant where he worked for almost 40 years. He walked to and from work every day and did all his errands and social activities within a one-mile radius of his home and the plant. Days off and holidays were spent with family or fixing up the house. Ron has never been on a plane, a train or a bus, and, although he was an experienced forklift and payloader operator in the open hearth, he never owned or drove a car. Delores worked at the co-op grocery store, also within walking distance of home. The view of the steel plant from his home on Victoria Road was an accepted part of Ron's life, as was the dangerous nature of the work. He still lives in the same house on Victoria Road, next to his two daughters who have also lived in the neighbourhood for most of their lives. He considers himself to have been blessed with a rich and full life.

Many workers were following a family tradition. Pat Wall's father, a Newfoundlander as well, was blacklisted in the 1923 strike. He was rehired by the company, then laid off without a pension in 1946 after 48 years of service. Pat worked for the company for 44 years. In 1987, he was "out the gate—what was left of me. I got a steel knee, a broken back and a lot of sad days gone."[14] Gordon Kiley is a small, wiry scrapper of a man in his early 70s, an amateur boxer, with fierce blue eyes. To this day, he is very bitter that his father became ill from working at the plant and was forced to retire without a pension. Gordon, who eventually became president of Local 1064, started work at the blast furnace in 1950, but was injured in 1963 and assigned to an "easier" job dumping slag out behind the plant.

Pat Wall worked the blast furnace. He recalls that whenever he opened the door to the furnace, so much smoke would cover him from the backdraft that his friends wouldn't recognize him as he walked home from the plant. Lorne McIntyre also worked the blast furnace and, later, the open hearth. He recalls how most of the open hearth workers were Catholics, like himself. In fact, the workers in the mechanical department and the machine shops—mostly Protestant and members of the Orange Lodge or the Masons—called the open hearth area "Little Vatican."

With a roar of laughter, Lorne recalls how these men would wear their religious medals to work. When they straightened up after bending over the 1,600° Celsius heat of the open hearth, he said, "you could smell the hair and flesh burning" as the medals settled back on their chests. He wonders how the young men of today with "rings in their ears and noses" would fare. Adds fellow worker Harry Muldoon, "The bricks were so hot, your feet were hot, your ears would burn, your pants would singe, the hair inside your nostrils would burn."

If water and steel were not mixed properly before the steel was tapped from the furnace, the whole 6,250-ton blast furnace could blow up. Many men lost their lives in this terrible way. With dark humour, generations of steelworkers have told the story of what happens to these men after death. As a matter of course, all open-hearth workers go to heaven. When they arrive before St. Peter at the pearly gates, all they have to say is: "Another open hearth worker reporting, sir. I have done my time in hell."

There was great danger in working with hot steel. Clyde Hoban and Don Puddicomb reminisce about learning to "catch steel" from the older men. This was the process of carrying steel billets from one set of rolls to the roughing stage—essentially, the men had to catch the white-hot rods and swing them around their bodies to the next stage of production. If the steel rod got out of control, it could wrap around the worker's leg and brand his body while cooling down. The only way to free him would be to cut the cold steel directly off his body. "We called them 'hot worms' but they looked more like a snake coming through the woods," says Clyde Hoban. "Sometimes it would fly right up to the ceiling. Only experience kept us alive." Adds Nelson Muise, "If you saw a worm, you found the quickest and easiest way out of there before it came down around you."

Everyone agrees that the jobs at the coke ovens were the worst, and no one is proud of the fact that this is where the Black and eastern European workers were usually stationed. Ed Johnson grew up in Whitney Pier's "Scotchtown" and would go on to have a distinguished union career as president of the Nova Scotia

Federation of Labour in 1959 and as executive assistant to the president of the Canadian Labour Congress in the 1980s. He says that the coke ovens were the part of the operation "closest to hell," and remembers plastering his face with sulphur salve to protect it from the heat. He would work by a big pond that simmered with a rainbow of chemicals in the summer, sending out fumes that could knock a man down. In winter, he worked ten hours a day on a coal bank in wet rain and snow, and walked the mile back to Whitney Pier through the polluted barrens.

Workers tell of flares overhead and fires underfoot. Their caps would burn off their heads while the thick wooden clogs they wore would catch fire and the skin would peel off their feet. Workers needed asbestos gloves to open the doors of the ovens, but the company did not supply them. Nor did it supply masks until the mid-1970s. Even then, they were poor quality and far inferior to the standard in the industry, such as the masks supplied to the workers at the Hamilton steel plant.

The company did supply rain gear, but the men could not wear it because the hot rubber would melt and burn their skin. On one particularly hot day, a worker was sent up onto a coke oven battery with a thermometer strapped to his leg, to record the temperature. It registered almost 150° Celsius before bursting while the worker was still standing on top of the battery.

Al Lewis worked as a by-products operator at the coke ovens. He measured the tanks containing benzol and other toxic chemicals, and supervised their loading onto the open cars for transportation off the site. His widow, Barb Lewis, recalls that he would walk knee-deep in chemicals regularly. One time, she remembers, benzene was leaking from a storage tank, and Al and other workers built a deep ditch around the spill to catch the runoff. There was a torrential rain that day, and the ditch overflowed. Al spent from one o'clock in the afternoon until after midnight "swimming in that toxic soup," turning off the valves.[15]

"You can't hide from the smell of the coke ovens," it was often said. While the coal was baking in the furnace, gases poured out of the battery, surrounding the workers. Many coke oven workers vomited every day at work and expelled a steady stream of black phlegm from their lungs and their noses. Ed Johnson recalls that people could always tell which men were trimming coal. They could wash it off, "but it would ooze out of your hands and eyes."[16]

Lunch—or dinner as they called the noon meal—was planned around the furnace. The sharing of food was an important ritual and one the men looked forward to every day. For many years, the company provided no lunchrooms and the men would cluster by skill or ethnic group wherever they could find a little space. For instance, the Polish and Ukrainian workers had a spot covered in corrugated iron behind one of the furnaces known as "Budapest." It was off limits to all but members of eastern European origin.

The men used the heat from the blast furnace or the open hearth to cook their food. Corned beef and cabbage or herring and potatoes would be boiled in a bucket; eggs, kippers and bacon would be fried on a shovel. Mussels would be cooked in an empty paint can. The men would put stew in a large spoon used for pouring iron and "cod-jig" the food up and down by a wire. During deer season, the workers always looked forward to the treat of deer steak, and the men swear that, more than once, they shared beaver roasts. Don Puddicomb loves to tell how the new employees would bring their cans of beans or spaghetti to work and stick them in the fire. The seasoned men knew that the heat would make sealed cans explode. But "of course, we didn't tell them."[17]

Clyde Hoban worked for a while in what he said was the dirtiest job in the plant—cleaning the inside of the towers where the company made mothballs from naphthalene. He said that the smell was so bad it would permeate his lunch bucket, making the food inedible. There was no washroom, or running water, and he had to heat his coffee on the steam hose and "blow the crud on my tea mug to the other side to take a sip." Why did he work at this job? Because "the money was clean at the end of the week."

The men were very aware of the hierarchy in the plant and the community. Lorne McIntyre remembers being told as a child that the steel mill was for people who couldn't cope in school: "If you were not a good student, they would turn your head and point to the smokestacks and say, 'Boy, that's where you're headed.'" He had six brothers who went into "the professions" and was hurt when someone asked him where he went wrong. He quickly adds, however, that if his brothers got one quarter of the satisfaction from their work that he got at the steel plant, they were blessed.

The men were never asked to give the company any advice on operations and were treated by the bosses like second-class citizens. In town, there were bars for the "big shots" and bars for the workers. Or, as George MacEachern would say, different bars for "people who worked with their hands and people who thought they were using their brains."

Lorne McIntyre studied to become an electrician, a job that had a higher standing in the plant and guaranteed him access to a lunchroom. The first day he came to work as an electrician after his training, he saw a foreman and a worker in his unit in a drunken brawl on the floor. Lorne realized that there wasn't such a big difference between the skilled and unskilled workers after all: "I says to myself, it's like the devil said when he went to the Senate—'It's good to be home again.'"

In the 1930s, in reaction to enormous pressure from the workers and the community, the company set up an on-site hospital, which operated until the 1960s, when local facilities improved enough to take over. The hospital had three levels with beds, an operating room, a treatment room, several doctors' offices, an X-ray room and a nursing station. A doctor was on duty weekdays and nurses—who had to be single and live in residence—were present around the clock. Because of the enormous number of plant accidents, the hospital was always busy. Records for 1953, for instance, show that 18,582 patients were treated that year.

Syd Slaven describes the first ambulance service as "crude." He recalls how the first driver in 1943 didn't know how to drive, so he was sent to town with a dollar, where he purchased his licence and reported to work the next day. The ambulance had no equipment except a fold-up canvas stretcher that was like a bag. "When the ambulance responded to a call," Slaven writes, "the victim was thrown in like a sack of flour. If an employee had a spinal injury that did not kill him on the job, there was a good chance the trip to the hospital would."[18]

The men remember the feeling of inferiority they had when they used the hospital. As one said, "You came in, cap in hand, if you had a hand left." But their irrepressible humour seldom left them. Everyone had a favourite story about a formidable head nurse they simply called "Murphy." One worker, who was in a long convalescence in the hospital after an accident at the plant, loved his rum, which was strictly forbidden. Murphy could smell liquor a mile away, so visitors would bring him bottles small enough to hide in a box of Kleenex. Every time she smelled rum on him, Murphy would search his room and bed. The day he was discharged from hospital, she stuck her head in his room and yelled, "Hey, Smith, don't forget to take your Kleenex!"

In an unspoken rule, workers did not describe and would never publicly discuss fatal accidents that took place at the plant, so that the families would never know the terrible details of their loved ones' suffering. Years later, when they are asked to describe some of these events, the room goes quiet and the men have to catch a breath before speaking.

One horrific accident took place in 1977. A coke ovens gas line in the battery basement was being switched to another system, and highly combustible gas was allowed to escape into the surrounding area. Joe Crane, a gas tester, smelled it. "She's gonna blow!" he yelled, but it was too late. A huge explosion rocked the whole coke ovens site and the ensuing fire raged for a week. Don Puddicomb was one of the first to arrive at the scene. He saw two men—brothers—sitting on a smokestack. "All they had on was their shoes," he recalls. "Their skin was hanging off their faces and only by their voices could I tell which one was which."

Seventeen men were severely burned. A few ambulances arrived to take them to the city hospital, but most had to be taken by other employees of the coke ovens. Donnie MacPherson took two victims in the back of his station wagon. Don Puddicomb took Joe Crane in his car, sitting up all the way. "Joe asked, 'What do I look like?' I said, 'You're just fine, Joe.' I couldn't tell him his face was gone." At the hospital, victims lay on stretchers and wet sheets while the staff tried to cope. Joe was so badly burned that the nurses put a heater on each side of him to keep him warm, and his bedclothes caught on fire. Another worker's screams filled the hospital as the doctors cut off the skin hanging down from his arms.

Although no one died at the time, many of the gas explosion workers were so badly burned that they never returned to work. Several men died of reduced lung capacity years later, and the fight for compensation raged for a decade.

Among those never to leave the hospital was Clarence Keller, a resident of Frederick Street, directly adjacent to the coke ovens. Clarence was 59 and about to take early retirement. Despite warnings at the hospital not to try to see her father, young Debbie Keller went to his room. She was shocked to see his head the size of a basketball. Years later, she recalls the horror of her father reliving the accident from his hospital bed—remembering the moment he realized that a small bubble of oxygen was holding back the explosion. "Leave it alone! Don't touch it!" she recalls him yelling for years after, whenever the morphine kicked in. Clarence is still alive, living in a long-term care home. He has had repeated strokes and has prostate cancer. He is totally blind.

Several years later, there was another terrible explosion, this time at the blast furnace. When molten iron poured out of the furnace mouth, all but one of the workers managed to get clear of the area. But Bennie Delorenzo, who was on the scene, remembers that one man, John Farr, was trapped in the slag pit. His two "buddies," Henry Gear and Jim Sheeves, went back to find him. When the three bodies were recovered, burned to a crisp, Sheeves and Gear had Farr's body cradled in their arms.

Jobs Before Life

For many years, the workers of Sydney Steel refused to recognize the health hazards associated with their jobs. Yes, cancer was prevalent, but cancer was part of life at the coke ovens. The workers would take leave for radiation treatment in Halifax and return to the ovens as soon as it was finished. Cancer was never recognized by the province or the company as a legitimate work-related malady. In fact, the men could claim compensation for only one condition: pneumoconiosis, or "black lung."

Most chemicals, such as the insulating and coolant by-products they used to rub on aching knees and elbows, were seen as benign. As Don Puddicomb explains, "We couldn't even spell pollution and no one had any idea what we were putting in the oven. You did what you were told. If you asked questions, you were told, 'Fill your shovel or fill your coat,' meaning you were fired."

As well, it was controversial within the union to raise environmental and health concerns too strongly. The company met any complaints or even questions about such concerns with a threat to close the plant, and the fear of unemployment kept the men in line. As one worker says simply, "When you have nothing, even something looks attractive." Donnie MacPherson, who worked as an electrician for almost 30 years, explains the attitude: "Those of us who worked at the coke ovens had always been vaguely aware that there were health hazards involved, but we lacked any real understanding of their scope. We knew little about the quantity and toxicity of the poisons being poured into our air, water and soil."[19]

It would take time and perspective for the workers to understand that an even more dangerous predator than accidents was stalking them, and it would be years before they would realize they had spent their whole working lives on one of the worst hazardous waste sites in North America. But cancer rates were soaring in the community and within their ranks, and it was clear the company was starting to notice. The workers remember the day in the early 1970s when men encased in full environmental protection suits arrived at the coke ovens site to take samples of several chemicals. No one from the company explained what these men were doing there or ever reported any findings to the union.

Ron Crawley Jr., now a researcher with the Canadian Union of

Public Employees in Ottawa, learned more about the danger the hard way. As a child growing up in the shadow of the plant, he would play in the fields and the contaminated barrens around the plant. The constant noise and smells from the plant were an accepted part of life that would only be noticed when family and friends from out of town would comment on them. Ron and his friends loved running into what they called "London Fog," when the coke ovens gases would be released and drift across the field between the coke ovens and his home. They would constantly be investigating the black and obnoxious-smelling tar brook and the tar pond into which it ran. It was not uncommon for a boy's sneaker to slip into the black water as his daring and curiosity brought him too close to its edge.

In 1973 at the age of 19, Ron went to work at the plant for the summer, but collapsed with severe abdominal pain. A local doctor told him he had ulcers, but on closer inspection, Ron discovered he had lymphoma. "I had to have a piece of my bowel removed and receive extensive radiation treatments," he remembers, the effects of which still haunt him to this day. After recuperating for a year, he tried to get back on the plant for the summer, but neither the company nor his family wanted him to return. Two years later, he left Sydney for good. Ron looks back on these years with mixed emotion. At one level, he knows he had an upbringing full of rich experiences as part of a tightly knit community. But Ron also remembers with much sadness friends and family who died too young. He also expresses great anger at the refusal of authorities to recognize the danger of what was clearly a contaminated site, and the almost total lack of progress in addressing the crisis in recent years.

Dan Yakimchuk was the first steelworker and union activist to sound the alarm in the 1970s, and was maligned by both his co-workers and the community for taking a public position. In 1988, Dave McLeod and Donnie MacPherson, who worked for the union counselling the families of cancer victims, released a study showing that 74 of 117 deaths of coke oven workers in the previous decade had been caused by cancer. They also showed that the level of cancer-causing emissions of PAHs*, including benzopyrenes (a yellow crystalline hydrocarbon found in coal tar) from the Sydney coke ovens was greater than those emanating from all fourteen coking operations at the Dofasco and Stelco steel plants in Hamilton combined.

Through the relentless advocacy of these men, the workers would eventually come to learn that they had by far the highest cancer rates of any similar site in Canada and six times the national average. Small wonder. The men were inhaling carcinogens equal to smoking 35 packages of cigarettes a day![20]

Jerry Moore, a 43-year veteran of the plant, can now reel off the names of colleagues who died of cancer: "Collie MacCormick, inoperable lung cancer, 37 years old. Joe Thompson. His whole body was covered in these gross lumps. He died at age 41. I can think of six people off the top of my head who have died of cancer since I left the coke ovens in 1988. Others have lost body parts." Moore's uncle broke his leg merely rolling over in bed. A tumour was found in the bone. A neighbour, Peter Roston, gives an interview about his years at the coke ovens through a hole in his throat. He lost his larynx to cancer.[21]

However, back on October 13, 1967—Black Friday—health was not the priority of the workers, including Jerry Moore. The Hawker-Siddeley announcement to close the plant devastated the community. Steelworkers and their families banded together with local businesses to reverse what they called "this ruthless and premeditated decision" to deprive the community of its livelihood. In a "Parade of Concern," 20,000 people marched through the streets of Sydney shouting "SOS—Save Our Steel!" and demanded that the provincial government step in. In December, yet another company, the Sydney Steel Corporation (SYSCO), was incorporated by the province to act as a Crown corporation.

* Polycyclic Aromatic Hydrocarbons (PAHs): A large group of carcinogenic chemicals. Unlike benzene, a carcinogenic aromatic hydrocarbon that has a single ring in its chemical structure, PAHs have multiple rings—hence "polycyclic" or "polynuclear," terms that are used interchangeably. They are volatile, that is, they volatize in the air, and are difficult to destroy.

The workers were elated and accepted a five-year wage freeze to show their goodwill. Their subsidization of the operation would cost them dearly; relative to the earnings of other workers in Sydney as well as of other steelworkers across Canada, their wages fell steadily over the next decade. Much to their dismay, safety and working conditions did not improve after the government assumed responsibility for the operation (the two horrific gas explosions described earlier took place under public ownership), and the company continued to deny that there were any problems.

This fact has led some to speculate that public ownership is no better than private for workers and their families. But as historian Joan Bishop points out, the government of Nova Scotia never really wanted to run the company, and although it poured a lot of money into SYSCO over the next two decades, it undertook none of the improvements that would have made for a safe, clean and productive modern plant. The community watched helplessly as the company made mistake after mistake. Instead of diversifying the corporation's product line, SYSCO stripped the plant, spun off the nail mill and tried to close the bar mill, an action that was temporarily prevented by strong union protest. In the end, the company was left producing only steel rails.

The government refused to put equity into the operation, insisting instead that SYSCO finance its own improvements. A piecemeal modernization plan left the plant with aging and unsafe blast furnaces, while the company set production quotas that pushed workers and aging equipment to the breaking point.[22] And, of course, no one in government or SYSCO's new management ever thought to ask the union for advice. The workers knew that it was "diversify or die," but the new company was as arrogant as the old one.

As a result, business was terrible, and the size of the workforce was steadily reduced from 3,000 in 1967 to about 700 in the mid-1980s. "The irony of the sharp decline in sales," wrote journalist Linden MacIntyre, "is that it was caused by a lack of steel to sell rather than by any softness in SYSCO's markets." Economist Tim O'Neill, now vice-president of the Bank of Montreal, said that with the world steel shortage at the time, SYSCO must have been the only steel plant in the world losing money. In 1974, Premier Gerald Regan called the modernization plan a "dead letter." Millions of dollars of taxpayers' money had been wasted by a government reluctant to play any role but that of offering incentives to private entrepreneurs.

The terrible price of keeping the plant open without rendering it safe for the workers and the community was the continued contamination of the surrounding area and its inhabitants. Dan Yakimchuk, who is now one of the leaders in the fight to clean up the site, marched "through the mud and muck of Black Friday" with his family and says that "it was the sorriest thing I ever did." He marched because he knew that if the plant closed, at least 30,000 working people who depended directly or indirectly on it would be out of work. But he also feared that the plant was doomed—a fear confirmed by an expert from the Steelworkers' Union in Toronto who visited the plant later that year and told the union that this "dilapidated old thing" would have to close sooner or later. Most important, Dan Yakimchuk knew deep in his heart that the steel plant and the coke ovens were the cause of the death all around him, and he was getting ready to break ranks with his fellow workers and say so.

The story of the men who worked the Sydney steel plant and coke ovens is a testimony to the human spirit. From the earliest days of struggle, they realized that theirs was a fight for justice. They took their union and their politics very seriously, and ran their union meetings with the kind of order and discipline one longs to see in the legislatures of the nation today. They were undervalued by almost everyone but their families and silenced by the power elites of their province and country.

Retired steelworker Walter E. MacKinnon hopes they won't be forgotten. "When future generations pass the location of the coke ovens, chemical plant, the coal bank and other landmarks of Sydney Steel, they will perhaps see ghosts. We should tell them now, do not be afraid. They are ghosts of good and remarkable men."[23]

They're dumpin' the slag over to the steel plant,
They're dumpin' the slag in the middle of the night.
They're dumpin' the slag over to the steel plant,
Go back to bed, mama, everything'll be alright.
—CJCB Song Contest

CHAPTER THREE
In the Shadow of the Valley

While the workers lived lives of a Hobbesian variety—nasty, brutish and short—their indomitable spirits rose above much misery. So too did the spirits of their families and the larger community around them.

Almost anywhere in Sydney, the towering presence of the steel plant dominates the physical, economic and social life of its residents. Like other steel towns, Sydney was built along class lines, with the areas for workers and management clearly marked. Stately homes with sweeping verandas still grace the tree-lined streets where the plant owners settled. Solid middle-class communities with schools and hospitals nearby housed professionals, civil servants and some highly skilled plant employees. The company forced the workers, including the early immigrants and most visible minorities, to settle in the worst areas, particularly Whitney Pier, on streets such as Frederick and Tupper and on Lingan Road. The Pier, as it is known, now has a population of approximately 6,500, about one-fifth the total population of Sydney, and a distinct culture and physical appearance clearly connected to its origins.

The Pier

Whitney Pier had an infamous reputation in its early days. The workers were housed in company shacks or substandard houses they built themselves. Drunken brawls were common and rivalry between ethnic groups often led to violence. A 1901 newspaper editorial decried the "pigs" and "urchins" running in the streets and said that the "dirty shacks" the workers lived in "should be razed to the ground." Amenities and services were not provided equally to the people of Sydney and Whitney Pier. Pier residents were constantly complaining about the lack of roads, housing, water and sewage treatment. At one town council meeting, the town engineer was accused of ignoring the Pier's street plan and randomly erecting houses wherever the whim took him.[1]

From the beginning, it was evident that the worst pollution from the plant stacks would be carried by the prevailing winds directly over Whitney Pier. A 1901 newspaper editorial advised that the "smoke nuisance" from the steel stacks should be concentrated in Whitney Pier, "where their smoke and roar will be no vexation, while their great outpouring of wealth will bring abundant commercial prosperity."[2] Donnie MacPherson, who grew up in Whitney Pier, laughs aloud when asked if the decision to build a community of workers downwind of the plant was connected to their class and ethnicity. "Of course!" he says. "That's why the Pier is here, dear."

Generations of Whitney Pier residents lived their entire lives under the orange "rain"—or ore-dust as they called it—that belched from plant stacks all day, every day. Sometimes ore-dust chemicals stripped paint off cars and houses, and falling particles pockmarked cars driving across the overpass. Tiny, gritty black specks called particulates sprinkled the furniture inside the houses, making it necessary to dust every day to keep homes clean. As well, the coke ovens sent out steady billowing clouds of poison gas. The stench of chemicals ruined food and invaded sleep.

In the early part of the century, before the overpass was built, workers had to cross the train tracks on foot to get to work. In 1912, the company carved out a "subway" to carry traffic under the tracks. It soon became notorious, however, because it had no sidewalk for pedestrians, and became flooded and impassable whenever it rained. In 1963, an overpass replaced the subway, somewhat improving travel between Whitney Pier and Sydney. But the easiest way for pedestrians to leave the Pier was still overland, and residents and workers alike would walk across the coke ovens site every day. Children would pick blueberries from the site in the summer, skate on the thin ice that covered the site in the winter and slosh through contaminated coke oven mud in the spring.

A Community Is Born

In spite of—or perhaps because of—its early harsh beginnings and the constant blights of poverty and pollution, Whitney Pier became a vibrant community, its residents bonding in adversity. The ethnic diversity that had been such a source of tension at first became a source of civic pride as each group's heritage enriched the whole. Whitney Pier became one of the first Canadian experiments in multiculturalism, a community greater than the sum of its parts.

Although racial differences and even racial tensions would remain, many residents of Whitney Pier came to be aware that they had more in common with one another than with the bosses on the "other side of the tracks." They knew they were perceived as "different" just for living in the Pier, and ethnic differences slowly took second place to class solidarity. As well, almost every family had lost members to the poison of their surroundings. With time, Whitney Pier became a kind of extended family, forged by the terrible common bonds of cancer and untimely death.

Scots, Lebanese, West Indians, Chinese, Jews, Italians, Ukrainians, and other ethnic groups created distinct neighbourhoods, and the business area teemed with a variety of ethnic foods and restaurants, particularly in the boom years between the two World Wars. In the 1930s and again in the 1950s, a number of co-op houses were erected, part of the early Antigonish Movement (which sought to bring greater economic independence and self-sufficiency to communities through co-operatives) and thanks to the work of Cape Bretoners Father Jimmy Tomkins and Father Moses Coady.[3] Brightly painted wood-frame houses with little verandas and vegetable gardens sprang up in the Pier. By the 1960s, children were no longer attending school based on their ethnic background (although separation by religion would remain), and schools such as the one for immigrant Coke Oven children called "Cokovia School" were a thing of the past.

Benevolent societies sprang up to take the place of poor or non-existent social services and company benefits. Each society was based on ethnic or religious background, and had its own distinct dress and often its own halls or lodges for meetings. The Daughters of Jacob Aid Society was a society of Jewish women whose objectives were the moral and spiritual welfare of the community. They worked alongside the St. Michael's Polish Benevolent Society, the Ukrainian Workers Benevolent Association, the Orange Lodge, the Knights of Columbus, the Protestant Society of United Fishermen, the Universal Negro Improvement Association and others to enrich the cultural life of their communities. Besides welcoming newcomers to Whitney Pier, the societies cared for the sick, ensured that burials were in the tradition of the faith and raised money to care for the poor.

As every ethnic group had its own place of worship, Whitney Pier had an unusually high number of churches for its population. Methodist, "Holy Roller," the Salvation Army, Anglican, Catholic, United, Presbyterian, Greek Orthodox, Russian Orthodox, African Orthodox, Jewish Orthodox and even African Methodist Episcopal congregations were active in the Pier in their time, and many still are. Some were a hybrid of a number of traditions. St. Philip's

African Orthodox Church was created by Blacks who rejected the white Anglican church, which allowed them to attend but forced them to sit at the back and refused them the right to be married there by the priest. A little black-and-white building on a quiet street, the church still serves its community today, although a good Sunday attendance is barely over a dozen worshippers.

By and large, the churches played a non-political role in Whitney Pier, acting instead as centres for cultural and spiritual activity and as sources of support through hard times. The Social Gospel movement, a Christian-based response to the upheavals of industrialization and immigration that was key to the development of social programs after the Second World War, was present to an extent in the Pier. Largely found in the Presbyterian and Methodist churches, it sponsored action for improved housing and working conditions, and had a limited involvement in the labour movement.

Other churches, particularly the Catholic church, used their power in the community to keep the workers from taking political stands or action, preaching the virtues of accepting hardship in this life for a reward in the next. Many a Catholic worker was dissuaded from joining the union by the church. Former union leader Gordon Kiley says the Catholic church was always looking for "reds under beds," and says that when he complained to his priest about the tough working conditions at the plant, his priest admonished him to "pray harder."

Dan Yakimchuk remembers a visit from his priest after he first became active in Local 1064. "If you stay with the union," said the priest, "you're going to go to hell." Yakimchuk remembers with delight his answer and the priest's shocked reaction: "The problem with going to hell as I see it, Father, is that you can't enjoy the fire for all the God-damned priests."[4]

As the community matured, the role of women developed and grew. There were relatively few women at the Pier in the early years; in fact, women and children accounted for less than one-quarter of the population at the turn of the century. Women either were wives and daughters of farmers and workers or provided sewing and laundry services to the steel plant's workforce. Prostitutes did ply their trade for a time, but were driven out of Sydney by the town fathers. Households would take in male boarders, and some women ran small hostels and corner stores. The demand for domestics in the wealthier areas of Sydney made the entry of West Indian Black immigrant women possible despite strict policies at the time against Black immigration. As a result, West Indian Black women were the first immigrant group to make up a large number of women in the Pier.[5]

Women began to arrive in Whitney Pier in large numbers after the First World War. Historian Elizabeth Beaton says that the arrival of women and children and their subsequent settlement marked the establishment of Whitney Pier as a community. Some women came in the hopes of finding a husband. As Beaton notes, "More than one Whitney Pier woman can tell of doing farmwork or in-service to 'earn money for passage and to build up a dowry' of clothes and linens, all in anticipation of marriage to a young man known only by a photo sent in the mail."[6] As hard as life was for women married to the workers of the steel plant and coke ovens with their poverty wages, life for unmarried or widowed women was almost intolerable. Many women, unable to find work to feed and house their children, lost them to the state.

Gradually, women began to take jobs as clerks, cashiers, nurses and teachers. For a brief time during the Second World War, 1,000 women, many of them from the Pier, were hired to replace the men from the steel plant who had enlisted. They were stratified on the job by ethnic background in exactly the same way as the men, and when the war was over, they were sent back home to make way again for the "real" workers. While the stories of the women of Whitney Pier have rarely been documented and are therefore harder to find than those of the male workers, it is clear that it was the women who upheld the traditions of their communities and created a social life for themselves and the hard-working men.

The People of the Pier

Clotilda Yakimchuk is a woman of the Pier. She was born in a small house on Tupper Street, right by the "subway" directly adjacent to the coke ovens site. Company scouts recruited her father, Arthur, who worked on the Panama Canal, to come to the Pier from Barbados in 1917, and her mother, Lillian, followed in 1918. Her father worked as a "gas producer" at the open hearth, shovelling coal 12 hours a day. When his supervisor retired, he was told not to bother applying for the job, but was instead ordered to train the white worker "they brought in off the street." He left the mill in protest and supported his family of five by delivering coal in a horse and buggy around the Pier and by teaching piano. Arthur and Lillian loved music, sang with the church choir and taught their children to excel in school.[7]

Clotilda's family house was in the worst possible location. She remembers not only the choking ore-dust showering down all the time, but also the steady plumes of dirty steam emanating from the coke ovens—steam that contained carcinogens, she has since learned. Every Monday, her mother would put the washing out to dry on lines and, often as not, she would have to "put it through the washing board again." Yes, the air smelled awful. "But you just didn't think about it," she says. "It was an operating steel plant. It provided a living for my brother, my uncle and my father."

From early childhood, Clotilda dreamed of becoming a nurse. In the early 1950s (when schools were still segregated in Halifax), she applied to the two main Sydney hospitals for acceptance into their nursing programs and was told she was not "nursing material." She applied to the less desirable Nova Scotia Psychiatric Hospital in Halifax and was accepted—the first Black nursing student in the hospital. Her nursing career, however, came to a temporary halt when she fell in love with and married a young Dalhousie law student from Grenada named Benson Douglas, and moved to the island with him where she lived for 14 years.

Although she was expected to live the leisured life of a lawyer's wife and raise her four children, Clotilda declared that she needed to do something more and became the director of nursing for the psychiatric hospital in Grenada. When her husband died, she moved her family back to Whitney Pier and her mother's house while she looked for a job. She soon found herself working at the Cape Breton Psychiatric Hospital in Sydney and excelled so that she was soon directing in-service training for nurses—the first Black woman to do so.

With a steady job and money in her pocket, Clotilda set out to look for a decent little apartment for her family outside Whitney Pier so that she could be closer to her job at the hospital. No one directly turned her down, but when she arrived at the door, she was regularly told by the owners that the place was rented or "not suitable." The owner of one apartment she particularly liked and could easily afford wouldn't rent it to her as it had "ghosts and queer things" wandering about. This was in 1972! Deeply angry, Clotilda called some prominent members of the Black community, including Winston Ruck, the first Black president of Local 1064, and some sympathetic local politicians and leaders from other communities and launched the Black Community Development Organization. Over the next few years, its members fought for better community housing, established programs to keep Black youth in school and set up a community radio program to promote Black culture and history.

But relations with local authorities became strained when Clotilda set up a cultural awareness event called "Think-In on Black Culture." The local police called in the riot squad, armed with semi-automatic guns. Clotilda protested to scrappy steelworker/union activist/alderman Dan Yakimchuk, who brought a successful motion to city council to have the riot squad called off and the use of these guns banned. This was to be the beginning of a remarkable political union and a passionate love story.

Dan Yakimchuk was born on September 29, 1929, on Henry Street in "Hunkey Town," not far from Clotilda's house. Dan says this date matters because it was the eve of the Depression and his family was so poor when he was born, "my mother didn't know whether to throw me out the window or jump out the window herself." His father, who didn't speak a word of English, fled the Bolshevik Revolution in Ukraine in 1917 for the Sydney steel plant, where he worked on the open hearth. He boarded with a Newfoundland family in the Pier and married the daughter of the house, a devout Christian—"a saint, really," says Dan—and produced 12 children. Dan says he doesn't remember ever having a childhood.

Although his mother valued education, especially for the girls because they might "make a bad marriage," Dan hated school almost from the day he started. The curriculum at Holy Redeemer Boys' School was stifling, the Sisters rigid and prone to violence and, worst of all, he was never allowed to question the status quo. In grade nine, he rebelled, and he and his mother were summoned to Sydney to see the superintendent of schools. They had to walk across the coke ovens, of course, and Dan remembers his mother's varicose veins aching all the way.

He also remembers that the superintendent did an unforgivable thing. In front of Dan's mother, he slapped Dan hard across the face. "I was furious," Dan recalled. "I says to him, 'Some day, I'm going to get on the school board and I'm going to screw those teachers.'" Years later, when Dan was a highly respected member of the local school board, the only teacher he had ever admired told him that Dan had been "twenty-five years ahead of his time."

Dan ran away to the circus, where he found the people "crooked in an honest way," which was more than he could say for many of the church and government officials he had met to date. He then went to sea and got involved with the Seamans' Union, a relationship that he kept up for years, even after he started work at the steel plant in 1949. In one of management's typical hiring practices, Dan was laid off just before he had worked the 21 days necessary to become a member of Local 1064. He was laid off another 25 times in his career with the company. The hardest time was in 1961, when 3,000 men were let go at once. By then, Dan had a wife and four children, and they were to experience the kind of poverty he had not seen since his childhood.

They moved into a former warehouse that was so cold, Dan

would stay up at night to watch the fire so the kids wouldn't freeze to death. "You want to know how cold it was?" he asks. "We had a picture of Our Lord over the kitchen table. It was so cold that when you looked at Him at night, He had His hands folded. But when you looked at Him in the morning, He had His hands over His ears." They put their underwear in the oven at night and picked blueberries and mushrooms to supplement the sandwich spread and home-made bread that became the family diet. Unemployment insurance day brought a box of cookies to last a week.

His mother, worried about the children, talked him into going to see the folks at welfare. "So I swallow my pride and, well, it's lucky I'm not in the pen," he recalls. "This skinny little son of a bitch asks me what I'm doing there. I tell him I just need a half-ton of coal. Wouldn't give it to me, and him from Glace Bay. You'd think I was a gangster. So I say, 'Boy, one day I'm going to be an alderman and I'm going to screw the social workers!' " When he was an alderman in the 1970s, Dan Yakimchuk brought a particular zeal to social assistance reform.

Thus Dan, like Clotilda, came to his political activism through personal experience. He helped to start the Unemployed Workers' Union, modelled on the one started during the Depression by George MacEachern, and worked closely with the Canadian Labour Congress's Ed Johnson. He became well-known in the community for his defence of the unemployed and his fight for better housing, and was a regular guest on a popular radio show called "Talkback." He would describe an issue to his daughters, who would draft letters to the editor under his name, and he became a lightning rod for company managers and local politicians.

Dan was elected to city council in 1972 where he fought for better schools, particularly in the Catholic system. He charged the church hierarchy with living in a comfortable pew paid for by the workers (the miners even had a dues check-off for the church), and pointed out that the schools in the system were little better than slums. These comments angered the local clergy, who vigorously defended the church's schools. "Tell me, Father," said Dan, "how come if the teachers are so good, some of us are so dumb?"

Clotilda and Dan continued to work closely. They were united in their determination to clean up the worst slum housing in Whitney Pier, especially on the streets directly adjacent to the coke ovens. In the 1970s, they successfully lobbied city council to declare Frederick and Tupper streets a "Green Zone"—an irony not lost on them now, considering what they have since learned about the extent of contamination in the area. In phases, the streets were to be cleared of inhabitants and no new homes were ever to be built on them again. Years later, Dan and Clotilda would ask some embarrassing questions as housing went up again on these "closed" streets.

Dan and Clotilda fell in love as Dan's marriage was ending. For Clotilda, it was a struggle, although she was deeply attracted to Dan from the first. "After all, few men, Black or white, can cope with an assertive Black woman. . . . Here I am with my Afro, telling young people to be proud of our heritage, and I am falling in love with a white man," she says. "White is beautiful!" quips Dan, a twinkle in his eye.

Dan and Clotilda were married in 1983, with all their children present. Although it was one of the rare mixed marriages in the Pier, their love was obvious to all who knew them and it soon ceased to be a matter of curiosity. "I'm so very happy that Dan has come into this part of my life," says Clotilda. "We have something very special. We're a good team."

Before retirement, Clotilda went on to a distinguished career in nursing, serving as the first Black president of the Registered Nurses' Association of Nova Scotia in 1986, and the first Black woman to be appointed to the board of governors of the University College of Cape Breton. She describes with pride being asked to give a public address in New Glasgow in 1987, 35 years after being refused service in a local restaurant. She says that finding a balance between being determined to take every opportunity that comes her way and maintaining an ability to care for others is the secret of survival at the Pier. "If there is a challenge, you don't run from it," she states. "You won't always succeed but the challenge is trying to work at it. My work has been one of not only helping myself but helping someone else. I know I am the richer for this."

Clotilda has been a great support to Dan in his fight to expose the health hazards of the plant and the coke ovens, a fight that has cost him friends and colleagues. He started to ask questions about the high rate of cancer among workers when he was first elected to city council. "It was commonly known that a man would be lucky to live long enough to see his first pension cheque at 65. I said to myself, 'How come doctors and lawyers and clergy all live to be in their 80s and we don't?' " Dan initiated a board of health sub-committee to study cancer rates at similar operations around the world, and with the help of a concerned local doctor and a plant foreman, started to dig up material others didn't want to see.

His agitation was not appreciated. He took his case for more study to the provincial government, where senior bureaucrats and politicians slept or talked through his presentation. He was booed at meetings of both Local 1064 and the provincial Labour Council, where he was a vice-president. "Are you out to close the whole plant?" co-workers demanded. He was called a Communist on "Talkback" and by former friends, and his children were taunted at school. At one point, he couldn't walk down the street without someone shouting at him.

But Dan Yakimchuk wouldn't back off. He set out to teach himself about the environmental and health effects of the tar ponds and the coke ovens, and is now, with Clotilda, a recognized community leader in the fight to clean up the contaminated site and bring justice to those workers and their families who have suffered so much. For Dan, the fight is deeply personal: he lost his mother, brother and sister to cancer, all while they were still relatively young. Fellow union members now acknowledge that Dan was right when he warned them years ago about the dangers of their workplace. "Dan was the canary in the coal mine," says lifelong friend Gordon Kiley. "We just didn't want to hear 'cause we wanted those damn jobs so bad."

Living and Dying

Peggy Brophy is another Whitney Pier woman. In fact, she has never moved away from Lingan Road where she was born in 1937, a mile downwind of the steel plant. Peggy French came from a devout Catholic family of ten children and attended Holy Redeemer Girls' School. At age 21, she married Harry Burt who grew up around the corner on Victoria Road and who worked at the plant like her father. They moved into a little house in her parents' back yard and, like her parents, had a big family—six children—to whom Peggy devoted her life. Peggy French didn't deserve the trial that lay ahead of her.[8]

Everything important in Peggy's life has been preceded by a dream. At age 37, Peggy had what she calls a "pre-mission." She went to her family doctor and, as she suspected, he told her she was ill—a rare form of cervical cancer—and that she needed radical treatment. He asked her what she wanted to know. "Nothing, Doctor," she said. "You do what you have to do and I will work on living through this." Peggy sat down with her children at the kitchen table that night and told them the situation and that she could die. "I want you all to stay together, always make sure you take your bath, wear clean socks and shorts, and keep the house clean," she said. Peggy's two oldest children had to leave high school to care for the four youngest while she and Harry left for Halifax.

Peggy had two radical radiation treatment sessions in addition to a punishing series of 15 cobalt sessions. Three radioactive steel balls were surgically placed inside her for two 48-hour treatments. She had to remain totally still and was left completely alone because the room was too dangerous to enter. The treatment was so intense that it reduced the size of her bowel to that of a baby's and gave her radiation poisoning that lasted for years. She was deathly sick, lying on her sofa at home unable to eat for months on end. Several times she hemorrhaged so badly that she almost died, and she was racked with diarrhea that plagues her to this day.

Peggy needed monthly treatment, but they couldn't afford for Harry to miss work and they couldn't leave their young children alone for long. So Harry would drive Peggy to Halifax during the night, wait while she had her treatment in the morning and drive directly back to the Pier without sleep. At one point, Peggy was under such stress that she had a bout of angina and ended up in the coronary unit of the same hospital. Off and on for years, she would inexplicably lose the use of her arms and legs.

In 1989, the year after the coke ovens were closed for good, Peggy got a call that she knew was coming. Harry, who had worked for years at the coke ovens, had had a heart attack. When he went into hospital, a "cloud" was found on his lung and he was diagnosed with lung cancer. This did not come as a surprise to her, for a dream had warned her of this disaster too. For two years at home, around the clock, she nursed Harry, who wouldn't let anyone else care for him. So well did she nurse him that he only entered the hospital in the last days of his life. They never let on to each other how ill he was, and on the night he died, he turned to her and said, "Everything is all right now."

Peggy locked herself away in her home, which she had seldom left anyway. She thought about the people in her life poisoned by the pollution in their midst: her father died of a heart attack at age 42, one brother died of a heart attack, another is a colon cancer survivor. Her sister-in-law died of cancer and, more recently, a sister (also a cancer survivor) and another brother had heart attacks only one day apart. "My grade eight class. I can still see them in the graduation picture," she says. "Judy Gillis, she was a nun. She died of cancer. There was Betty Burke, she had cancer at the same time I did. There was the Harrigan girl who had cancer. There was a Cameron girl that had cancer. Also Joan Fraser. They all died. There was the Suter girl—she died of a heart attack. And more." She remembered how, as children, they would all turn their laughing faces up into the orange dust falling from the sky on days when it was particularly heavy to see who could keep their eyes open the longest.

On Christmas Day 1997, Peggy Burt had a massive heart attack and landed back in hospital. On Boxing Day, she received a visit from Eric Brophy, a childhood friend. The visit would change her luck and her life.

Eric grew up directly across the street from Peggy and remembers thinking she was a lovely girl, but at five years his junior, "too young." He married Lorraine Campbell, another girl from Lingan Road, after joining the air force where he served for 32 years. When he retired in 1982, he and Lorraine and one of their two children returned to Whitney Pier. He immediately was struck with the high incidence of cancer deaths among friends and family. "You'd hear they were out shovelling snow and then that night they would die of a heart attack. So everyone would say he shouldn't have been shovelling snow. I thought, 'Shovelling snow be damned. It's growing up here.' "

Eric attended funeral after funeral and started to question publicly the relation between these deaths and the emissions from the plant. He became a one-man neighbourhood support system to families in crisis. He walked friends with cancer around the block, bought their groceries and shovelled their walks. He counselled their wives when his friends died. Then Lorraine was diagnosed with a rare form of blood cancer and fell very ill. She lived for another ten years, but was always exhausted, and Eric became a full time caregiver.

Eric is clear in his understanding that hers was no isolated death. He rhymes off the losses in their immediate families. Lorraine's father died at age 58 of a heart attack. Her mother lost both breasts to cancer as did two of her aunts. Her brother has colon cancer. His father, who was a bricklayer at the plant, died of cancer. His brother, a bricklayer as well, has colon cancer. "Look down this street," he says, pointing to Lingan Road. "Home after home after home has been hit by cancer and heart disease. Not just one, but every single house, in a line. This house, the parents died of cancer, left it to the daughter. She died of cancer. Her husband died of a heart attack.

"Next door to that house, one of my best friends from school died of cancer; his wife had two kidney transplants and died. Next house, the daughter is 16 and has just been granted a trip to Disneyland with the Children's Wish Foundation. Next house, lady has Parkinson's, husband diagnosed with cancer. In the next house is a cousin of mine. He has cancer. The woman across the street died of cancer and her husband died of massive heart failure. Next house, another friend, another funeral; Frankie Bennett, he died. Grew up right on Railroad Street, under the stacks. And that's just one street. Every single street has the same story. You know what they have in common? They all grew up in Whitney Pier. The only thing that changes around here are the names in the obituaries."

Eric remembers that the school he attended as a child was located just under the stacks, and in the summer the windows would be open all day. He remembers drivers having to put their headlights on in broad daylight to see through the ore-dust some days because the visibility was so bad. Later, he learned that 70 per cent of the time, the winds carried all the pollution of the plant and the ovens directly over the Pier, and that the company never bothered to build the kind of tall stacks found in many other steel towns.

"As kids, we snared rabbits from the coke ovens and brought them home to our mothers to cook. We ate chickens that fed on the coke oven grass, and sucked on coke oven ice chips when we played hockey in the winter. Our mothers all grew 'victory gardens' in the war and we ate vegetables grown in soil covered in ore-dust. You want to talk 'pathways'?" he says, in reference to the scientific term describing all the ways pollution enters the human system. "We had pathways all right. They were everywhere. But the government told us it was our unhealthy 'lifestyles.' What a pack of lies. They just don't want to accept the blame and open a flood of liability cases. I see the hurt and I see the suffering and I am angry."

When Eric heard that Peggy was in the hospital, he knew what she had gone through and wanted to comfort her. They were both lonely and bruised from the events of their recent lives. Peggy, deeply spiritual, would pray, "If there is somebody out there, send them along." One night, about a month before her heart attack, Peggy saw Eric in her sleep at the foot of her bed, telling her everything would be all right. During those same months, Eric would walk his dog around the neighbourhood and also pray, "If there is someone out there who needs me, please lead the way."

When he entered the hospital room with flowers, he knew he could stop praying. "Not a pretty sight, eh?" Peggy says now, referring to how she looked. "Oh, gorgeous," he counters. "Just gorgeous." Peggy asked the nurses if they could have five minutes alone. Eric noticed that the nurses, who were watching her heart monitor closely, were chuckling out loud. Peggy's happiness and excitement were showing up on the screen and the nurses gave them another five minutes alone. "When I left, I said, 'I'll be back.' I've been visiting ever since," Eric says.

Peggy and Eric were married in August 1998 with all their children present. The bride was radiant. Their cozy home is filled with the smell of baking bread, the walls covered with pictures of beloved family. "To me, finding Eric was everything," Peggy states. "He brought me out of the house. He gave me a whole new life to live." They both know that their time may be limited. Peggy says that she wouldn't know what a day without physical pain would feel like. But she believes that there is good in everything and that there is always hope. Finding Eric is proof of that.

Eric has emerged as one of the leaders in the fight for justice for the people of Whitney Pier. "I look at what went on here over the years, the cancer and the heart disease, and I wonder how much more the people of this community can endure," he says. "We have a hell of a mess here, even though some people still don't want to face it. When I look at genetic mutation, my fear is for future generations. The stacks have closed, but the problem is still here. Well, so am I."

A Mi'kmaq Woman

Shirley Christmas did not grow up in the Pier, but she knows all about the kind of discrimination so many Pier residents have suffered. Shirley is a Mi'kmaq woman, a daughter of Membertou, the Aboriginal reserve in Sydney that her people were banished to in the 1920s when the province finally succeeded in forcing them off their rightful lands on Kings Road. Shirley Christmas, who is now a prominent poet and spokesperson for Mi'kmaq rights, is a direct descendant of Ben Christmas, who was the band's chief during the 40-year relocation battle.

The 40-year campaign to force the Mi'kmaq people off their Kings Road reserve was launched in 1877 by a prominent Sydney lawyer, Joseph Gillies, who felt that they were driving down the value of his newly acquired property next door. Gillies argued that "Indians and Negroes were an undesirable class and an embarrassment to the citizens of Sydney" and that the Mi'kmaq were annoying him "to a point really beyond endurance." Gillies crusaded tirelessly against the tiny band, but was not successful until he won an amendment to the Indian Act in 1911. The Exchequer Court was given the right to override the band's objections to a forced move because the reserve was located in a community of over 8,000, and therefore the larger community's wishes had a right to prevail. Said Frank Oliver, federal minister of the interior at the time, it was unfair that the "right of the Indian should be allowed to become a wrong to the white man."[9]

In 1915, the matter came before Judge Audette of the Exchequer Court, who heard 34 witnesses, almost all of them white. Gillies filed his own statement: "Their habits and habitations are filthy and unsanitary; they do not and cannot assimilate with the white people and white people avoid proximity to them; they are addicted to debauchery, are quarrelsome and destroy the peace of the neighbourhood and are immoral." He noted that the only other local area that was as bad was Whitney Pier, "an area occupied by a foreign element."

Several witnesses, including Indian agent Cecil Sparrow, strongly disagreed in their testimony, citing the industriousness and law-abiding nature of the Mi'kmaq, but Judge Audette ruled that it was "expedient" to remove the Kings Road reserve: "No one cares to live in the immediate vicinity of the Indians. The removal would make property in that neighbourhood more valuable for assessment purposes and it is no doubt an anomaly to have this Indian reserve in almost the centre of the city or on one of its principal thoroughfares. The overwhelming weight of the evidence is to the effect that the reserve retards and is a clog in the development of that part of the city."[10]

The city fathers set out to find a "suitable" location and strongly favoured a site on Lingan Road, right where Peggy and Eric Brophy now live. It was owned by none other than Joseph Gillies who sold the property to the government out of what he proclaimed was the goodness of his heart. The site would be perfect for the Mi'kmaq, he assured authorities in Ottawa: "They would be near their church as I have stated, near the world where they would be employed and yet they would be distinctly separate from white people and from all temptations." Gillies insisted that being near the emissions from the coke ovens would pose "no danger whatever."

But the Mi'kmaq were not co-operating. In 1920, Duncan Campbell Scott, Deputy Superintendent General for Indian Affairs in Ottawa (and a beloved poet for generations of Canadian school children) had been granted $20,000 by Parliament to be used in "removing these Indians." He reacted with anger to their demands for a decent site: "The Indians, I might say, are still as unreasonable as they have always been in the past, in this connection. I met them, at the schoolhouse, to discuss the matter with them and learn their desires. They finally decided on a location further out the Kings Road. . . . The location they ask for could not be considered as it is in the direct line of the present and future growth of the best residential section of the city."[11]

In 1926, the government purchased what journalist Geoffrey York calls "a worthless plot of swamp, rocks, and woodland" on the southern fringe of Sydney for $5,000 and forced the Mi'kmaq to transfer there, razing their former homes even as they were moving out. The government had picked as cheap a location as it could find to become the Membertou reserve. In a 1925 letter, Nova Scotia Indian Superintendent A. J. Boyd wrote, "In locating the families, it is considered desirable that they be so located as to render necessary the minimum of grading of roads and ditching."[12] Indeed, the community was not connected to the Sydney waterlines until the mid-1950s, and because the area was a poorly drained swamp, the modest bungalows were often flooded and children became chronically ill from the dampness.

But most distressingly, the new site had no access to the water. For a maritime people who depended on fishing and canoeing for their livelihoods, it was the rape of a culture. Said Bernie Francis of the Membertou band: "Water is a vital element of Mi'kmaq culture. Mi'kmaqs in Membertou have difficulty exercising their customs now because they are no longer near water. The feeling of being close to the water was taken away from the people. It's like being chained."[13] As a result of losing both their traditional livelihood and the opportunity to live near the business hub of the city, 90 per cent of the Mi'kmaq of Membertou today are unemployed or severely underemployed, a rate unchanged for many years.

It is here that Shirley Christmas grew up. One of 15 children, she was born in 1950, her mother a Maliseet of New Brunswick, her father a Mi'kmaq of Membertou. She remembers, as a child, reading such terrible things about her people that she would declare, "God, I'm glad I'm not an Indian!"[14]

When she was five years old, Shirley was sent to the residential school run by the Catholic church in Shubenacadie, northeast of Halifax. Shubenacadie and Eskasoni, southwest of Sydney on Bras d'Or Lake, were the two sites chosen by the Nova Scotia government in the 1940s for an Indian "centralization" plan. All of the province's Mi'kmaqs were to be crammed into these two reserves in order to save taxpayers' money and to make them—in the words of J.A. MacLean, the federal Indian agent at Eskasoni—"decent chaps," maybe even "one of the greatest tourist attractions the province has to offer."[15]

Thousands of Mi'kmaq were forcibly uprooted from their home reserves and moved to one of the two designated sites; their former houses, farms and schools were burned to the ground. Although Chief Ben Christmas successfully resisted the relocation of the residents of Membertou (for which he was called a "left winger" by the Indian agent), many of the band's children were forced to attend the Shubenacadie and Eskasoni schools nonetheless.

The goal of the schools was to assimilate the Mi'kmaq children into white society by training them to become farmhands and maids. Shirley's school day consisted of only two hours of class; the rest was spent in manual labour. "I spent more time scrubbing floors with a toothbrush than I did learning the King's English," she recalls. She did not see her parents from September to June, and was only allowed to see her beloved older brother, attending the boys' section of the same school, for 15 minutes a month. If she spoke to him as they passed in the hall, they were both punished.

Two sisters, one older and one younger, attended the residential school with her. The older sister contracted tuberculosis and was sent home. The younger sister, only five years old when she first arrived, was a bedwetter. Every morning, the girls had to make their beds so wrinkle free that they would pass the nuns' gruelling inspection. (This task was part of their training for their future vocations, of course, and one so well learned that to this day Shirley has to force herself to leave some wrinkles in her bed when she makes it up.) Shirley's little sister always gave her "sin" away with a trembling lip. The nuns would strip the bed in front of the other girls, and if the sheets were wet, they would push the little girl's face down in them and then make her wear the offending sheets around her head like a turban for the rest of the day.

Shirley was strapped simply for being the older sister—ten strokes each to her face, hands, legs and buttocks—but when she was caught changing her sister's wet sheets during the night, she was given the worst punishment. For days at a time she was locked without food in a small pitch-black closet, a terror that would haunt her dreams until the school burned down many years later. When she was 12, Shirley told her mother that she would die if she was sent back to that school. She drifted in and out of local schools and jobs and moved to Toronto in her late teens, happy to leave her sad childhood and Membertou behind.

One summer night in 1973, when she was working the night-shift as a maid at the Royal York Hotel in Toronto, Shirley Christmas heard the loud, steady, unmistakable beat of drums. She had not grown up with the drums, so this was not a familiar sound. But she knew that the songs were very old and sacred. Terrified, she searched for the source of the sound. She called home to Membertou; her father had just died. Shirley knew it was time to go home.

In the mid-1980s, now a mother of six, Shirley started to write. She had set out upon a journey to connect herself to her heritage through traditional dance, music and sweat lodges, and the drums in her head, which she was hearing again, were calling to her to tell the story of her people. She was learning about the years of violent discrimination and dispossession just when another Membertou native, Donald Marshall, was unjustly convicted of a murder he did not commit. The racist overtones of that trial deeply affected her, as they did many others in the community, given that the man convicted was Aboriginal and the man killed was Black. Dan Yakimchuk remembers with fresh anger that when he remarked to a police commissioner that the evidence against Marshall was pretty flimsy, the commissioner said that it didn't matter because "we got rid of a nigger and an Indian at the same time."

Shirley came to understand that her ancestors had chosen her to impart their pain and anger to the community, and that she would do this through her writing. When her first volume of poetry was published, the Membertou elders gave her a Mi'kmaq name—Kiju Kawi, which means "Mother Quill, She Who Speaks the Truth."

Shirley's journey brought her face to face with another part of her history that she had tried to ignore: what had become of her people's ancient summer camping grounds? She remembers, as a child, that she never saw the sun for the ore-dust, and it wasn't until the plant closed in 1988 that she first saw blue sky in Sydney. But like most Sydney residents, she never questioned the pollution, and like most Mi'kmaqs, she had long ago lost touch with this sacred place. She had forgotten that the steel mill and the tar ponds were built on traditional Mi'kmaq lands.

A friend took her to one of the early government briefings on the toxic site, and what she heard made her realize that she could no longer run from this issue. She was so upset by government inaction that she wrote to Prime Minister Jean Chrétien, who had recently played golf at a nearby resort, to invite him to come to

Sydney and play a round of golf on the tar ponds. She never received an answer.

Her first visit to the site was very painful. "I felt so much pain and anger that first time, I could hardly bear it," she recalls. "My ancestors were there, I felt them everywhere, going back to a time before the settlers. They were hurting, crying out to me to apologize to Mother Earth for this dishonour." She held a cleansing ceremony and brought offerings of sweetgrass and tobacco so that the Creator could hear her prayers.

That night, she had a terrible dream of being chased to the top of the tallest building in Sydney, where she looked out over what seemed to be a war zone. Everything was in flames and frantic people were running about in terror. Out of the harbour came a huge snake-like monster made of mercury and arsenic with flames streaming out of its mouth. Then she saw an elder talking to a little boy and the two of them searching the sky for eagles. They started to chant and the eagles began to circle, dropping their feathers over the terrified people below. She realized that the Creator was giving them a gift of healing with these feathers, a chance for the people to redeem themselves by giving this gift back to nature. She picked up a pure white feather and vowed to the ancestors that she would use her gift of language to atone for humanity's desecration of the land and water.

In 1996, Shirley Christmas addressed a public forum on the tar ponds cleanup.

Long before the birth of my great-great-grandparents, my ancestors lived on this land. They believed that the Earth was their mother. For generations and generations, my people cared for and respected this fine lady that produced life for all living things. In return, she provided care and protection. My ancestors saw the beauty of this land, breathed the air of sweetness, drank the waters of purity, ate the food of the land, and fished from the seas of abundance. How fortunate my ancestors were to have seen the beauty of Muggah Creek back then!

Times have changed since then. From across the country, the land and my forefathers suffered immensely. Today there is no longer balance. There

> is only 700,000 tons of toxic waste that covers the mother of all life, our Earth. Today, the air has no sweet aroma. The water I drink is no longer pure. The lands and seas are emptied of thousands of species of life. Those that are left are contaminated with chemicals I can't even name.
>
> If ever I hurt before, it can't hurt as much as I do right now. I ache with burning pain and anger; anger because yesterday, there was beauty. Today, I see the ugliness of greed slowly devouring what remains of my mother, the Earth.[16]

Recently, Shirley had another dream. She was walking with a young child along the shores of Muggah Creek on a lovely summer day. Everywhere she looked, there was breathtaking beauty: trees and flowers in full bloom, gulls circling the clear blue sky, herons and sandpipers fishing abundant waters. The air was as sweet as the day of Creation. Shirley came upon a sign posted on a gate. It read: "This used to be the Sydney tar ponds." For one instant, Shirley had the flash of an ancient terrible memory. She had been in this place before, when it had looked very different. But the memory quickly faded and was replaced by an overwhelming sense of serene peace. This was a day of atonement. Her ancestors were watching. She was reborn.

Let's save our industry
Let's save our industry
Let's save our industry
The industry we need.

It's brought us joy and brought us tears
It's been here over sixty years
It built our homes and stilled our fears
And made this island what it is.

—Song written for the Parade of Concern,
November 20, 1967. Lyrics by Charlie MacKinnon

CHAPTER FOUR
State-Sponsored Crime

As the community lived through the mixed blessing of having the steel plant as an ever-oppressive neighbour, and as the workforce unionized and began to question the links between their illnesses and the mill's effluent, the government continued to put steel ahead of health. Making steel at any cost became an obsession.

From its earliest days, the steel plant was intimately connected to the public purse and to federal and Nova Scotia politics. In its 66-year life, there was never a time when grants and subsidies from all levels of government were not available when the company called for help. But this constant government support was just the warm-up act for the state-controlled phase of the steel plant saga. In the 30 years following the government takeover of the mill in 1967, over $2 billion in additional subsidies would go into the Sydney plant.

SYSCO

In December 1967, when the province of Nova Scotia took control of the failing DOSCO operations and created the Sydney Steel Corporation, or SYSCO, government bailout was seen as the only way to save the plant. Only a few short years earlier, the federal government had taken over the island's coal industry with the creation of the Cape Breton Development Corporation (DEVCO).

R.B. Cameron was appointed first chairman of SYSCO's board, and the plant immediately began to benefit from federal subsidies. Millions of dollars in unpaid loans that the steel plant owed to the federal treasury were written off. SYSCO also benefited from a more indirect federal subsidy: Cameron arranged to sell the coke ovens facility to DEVCO, the federal Crown corporation. DEVCO paid SYSCO $4.3 million, and committed to selling coke to SYSCO for five years at the guaranteed bargain basement price of $18 a ton, well below the costs of production.

The province came through for the new state baby as well. By October 1968, the Nova Scotia government of Premier G.I. Smith announced a $50 million modernization package—$20 million in 1970, with $10 million a year for the next three years. Production increased as the company secured new orders for steel rails from Korea, Chile and the United States.

No one seemed to even consider the need for the $6 million sintering plant that would have reduced air pollution. Instead, as R.B. Cameron drove production levels up to a 1969 peak of one million tons, air pollution worsened.

In fact, every type of steel mill pollution increased once the government took over running the plant. No pollution control measures of any type were employed—nor were they ever seriously discussed.

Government as Polluter and Regulator

Between 1967 and 1970, total dust and particulate dumping in Sydney increased by 71 per cent. In January 1970, an unbelievable

521.3 tons of dust fell on every square mile of Sydney![1] Federal Health and Welfare experts in the air quality division reported that sulphur dioxide rates had also increased 36 per cent in 1970. They warned that air quality "can be expected to deteriorate rapidly as steel production increases."[2] The report was classified as a restricted document and not released. SYSCO also continued dumping cadmium waste from the steel mill. Even though a plant in Quebec could have disposed of the waste properly for only $40 a ton, SYSCO dug a huge hole in the slag piles at the mouth of the estuary and dumped the waste there.[3]

Naturally enough, once DEVCO was running the coke ovens, it reverted to using its own poor quality coal. With a commitment to deliver coke at $18 a ton, the company never even considered other options. Whereas DOSCO had switched to higher-quality, lower-pollution imported coal, now DEVCO shifted the mix from 80 per cent imported, and 20 per cent Cape Breton coal back to 75 per cent Cape Breton coal and only 25 per cent imported.

The results were predictable. Poor quality coal increased pollution. Coal near the doors of the ovens received less heat than the rest of the baked coal, and rather than volatizing as gas, it created a tarry sludge that remained liquid. This substance routinely ran from the ovens to the ground, drained by surface water to Coke Ovens Brook and Muggah Creek.

By this time, Muggah Creek had ceased to look, smell or act like a body of water. Children, now grown, recall playing along its banks and marvelling at the progress of a flat stone thrown into the tarry mess. For a moment, it would lie suspended on the surface, and then be swallowed up by a liquid mass the consistency of chocolate pudding.[4] Although the estuary was not a contained body of water, it was universally known as the tar ponds.

The pollution created by DEVCO management was worse than it had been at any time since 1903. In the past, when dirty, Cape Breton coal had been used, there had been a full side-industry of by-product manufacture. While these operations could hardly have been considered pollution control, they at least diverted tons of toxic substances from direct dumping. But this side-industry had closed. When DOSCO began using purer imported coal, the supply of toxic waste available to the Domtar plant and DOSCO's own by-product operations had dried up. Domtar had left in 1962, and the remaining apparatus for trapping and processing coking waste had fallen into disrepair.

DEVCO made no effort to restart the by-product component of the coke ovens operations. In fact, it turned away American investors who wanted to buy the effluent moving down Coke Ovens Brook to make commercial by-products.[5] Instead, massive quantities of toxic sludge containing PAHs, cyanide, ammonia and phenols were dumped onto the ground and into the brook. For every 17 tons of coal, a coke oven would produce only 11 tons of coke, with the 6 tons in the difference released as coking gases and tars. With 117 ovens working in this period on a 17-hour coking cycle, the coke ovens were designed to process 2,735 tons of coal every day. That translated into approximately 330 tons of toxic waste being dumped into the environment, *from the coking operations alone*, every day.

The brook was so heavily contaminated that on a hot summer day in 1972 it actually caught fire and burned down a railway bridge. DEVCO workers were banging in bolts that held the enormous timbers of the company's bridge together when a spark hit the brook. It ignited a fire that spread so quickly, the bridge was destroyed before the fire could be extinguished.[6] No one thought the incident sufficiently unusual to report in the daily press.

As well as the noxious liquid emissions from coking and quenching, there was also a huge increase in airborne pollution from the coke ovens. A study by Environment Canada in 1973 advised that, in order to meet National Ambient Air Quality Objectives, the level of particulates, the coal dust and the coking dust, would have to be reduced by 98 per cent. Sulphur dioxide levels would have to be reduced by 54 per cent.[7] "Calculations of ground-level concentrations show that the maximum limit for particulates can be exceeded by a factor as high as 60 times," the report stated.[8]

The report also raised concerns that the slag waste from the blast furnace operations, dumped directly into pits along Muggah Creek, was being loaded onto trucks and "hauled away to be used as landfill."[9] Despite strong language about the nature of the concerns, the report recommended only that "consideration be given to adoption of the best practicable technology. . . . further negotiations should take place with the Province of Nova Scotia and SYSCO. . . ."[10]

The outrageous pollution from the coke ovens ceased to be DEVCO's problem when, in 1974, the federal Crown corporation sold the coke ovens back to the provincial Crown corporation SYSCO for $10 million. As SYSCO had $10 million in claims against DEVCO at the time of the sale, alleging that the coke supplied had been inadequate, the deal took place without a penny changing hands. However, over the five years that DEVCO managed the coking operations, it has been estimated that the federal subsidies from the arrangement, direct and indirect, amounted to as much as $40 million.[11]

In 1974, Environment Canada followed up on its assessment of Sydney air quality. Air pollution from coking operations produced emissions 2,800 to 6,000 per cent higher than the standard.[12] The report noted that "there are no air pollution control systems operating on the coke batteries. . . . Overall particulate emissions from the coke-making facilities. . . . amounted to 5,680 lb/day in 1972 and are estimated at 11,560 lb/day in 1975."[13]

All of the air quality reports from Environment Canada, like the federal Health and Welfare reports from 1970, were restricted and suppressed. Neither the workers nor the community had access to them until the late 1980s. And even then, it was not the government that released the data, but an enterprising electrician.

By chance in 1987, Donnie MacPherson was called in to replace the lighting in a SYSCO manager's office. While a co-worker watched the door, Donnie had a look in the file drawers for any evidence of pollution and health effects to buttress the work of their newly formed group, Coke Ovens Workers United for Justice. Quickly riffling through the files, Donnie reached a folder marked "confidential." Inside were the Environment Canada reports on air quality from 1973 and 1974. Taking the whole file, Donnie left the office quickly. When he had a moment to open and read the restricted reports, he was furious that the government knew how polluted the air was in Sydney and kept it a secret for so long.

Donnie MacPherson released the reports to the media and, while not saying how they came into his hands, explained that they had been in SYSCO's files. Company spokesman Harvey MacLeod never accused him of theft. Rather than admit the truth—that the provincial Crown corporation had refused to adopt pollution abatement measures and suppressed pollution information for years—Harvey MacLeod denied to the media that the company had ever possessed or seen the files.

In 1976, the increase in air pollution attracted the attention of the federal department of Health and Welfare. It conducted a study of children in the Sydney area to see whether increased respiratory problems were associated with the poisoned air. This study was fascinating in two ways. First, and not surprisingly, it found an association between children's breathing difficulties and bad air. The department's report concluded that air pollution limits "cannot be exceeded, as it has been the case in Sydney for some time, without having some measurable negative effect on respiratory function."[14]

Second, the study's air measurements raised another key issue, although the report did not dwell on it: equity. Quite simply, not all Sydney children were polluted equally. The prevailing winds meant that while every part of Sydney was polluted some of the time, the community of Whitney Pier was polluted nearly all the time.

The mean exposure to particulates in Whitney Pier between July and September 1975, for example, was 125.3 micrograms per cubic metre (Mg/m^3), with the highest reading 292 Mg/m^3. On the other side of the tracks, along Kings Road near St. Rita's Hospital, the mean was 39.6 Mg/m^3 with the highest reading 59 Mg/m^3.[15] In relative terms, the worst day for bad air in the nicer part of town was a clear day in Whitney Pier. Residents of Frederick Street and nearby Pier neighbourhoods received the drainage of sludge from the coke ovens site, as did residents of Ashby Street along the other side of the ovens, where Ron Crawley's family lived. In addition, they experienced the highest occupational exposure. As most workers lived in or near the plant, the prevailing winds brought pollution in the form of dust, ash and poisons over their homes and into their lungs nearly constantly.

It did not seem to matter how much evidence was amassed by federal departments responsible for the environment and human health. No alarm bells rang anywhere—until, that is, the Fisheries department noted that a crustacean was in trouble.

Enter the Lobster

In 1980, the federal department of Fisheries conducted surveys of Sydney Harbour lobsters, testing them for contamination with PAHs, the deadly, cancer-causing chemicals. Over the years, the coking operations had been creating and dumping enormous quantities of PAHs and other aromatic hydrocarbons such as naphthal, naphthalene, benzol, benzene, pyrene and toluol, and their related compounds. The steady migration of chemical sludge from Coke Ovens Brook to Muggah Creek did not end at the mouth of the estuary, of course—it continued to the harbour and the open sea beyond.

Conclusive evidence came in the form of PAH-contaminated lobsters. Lobsters in Sydney Harbour also had high levels of PCBs, mercury, cadmium and lead. Indeed, they had the unenviable distinction of being 26 times more contaminated than lobsters from the famously polluted and much larger Boston Harbour.[16] Subsequent testing of the tidal flats along Muggah Creek revealed the once teeming ecosystem to be "completely devoid of all life," with water that was "acutely toxic to test organisms."[17] In 1984, studies confirmed that the lobsters were heavily contaminated, as were mussels collected in the harbour.[18]

The department of Fisheries took the first step aimed at

protecting public health in over 80 years of Sydney steelmaking. In 1982, it closed the lobster fishery in the harbour's south arm. On one side of an imaginary line through the harbour, lobsters were safe to trap and eat; on the other, they were not.

Testing by Environment Canada pointed to the obvious source contaminating the fishery: SYSCO's operations. The federal department estimated that the daily discharge into the harbour was "735.5 pounds of phenol, 10,447 pounds of ammonia, 919.4 pounds of cyanide, 2,058 pounds of thiocyanate and 995 pounds of oil and grease," representing pollution rates *ten times* the industry average.[19]

Meanwhile, the source of the pollution continued its business as usual. SYSCO had capitalized on its environmental irresponsibility by demanding additional federal funds, arguing that further modernization could reduce pollution. In 1981, the federal government committed to a new $96 million modernization plan, of which $14 million would be earmarked for pollution control in the blast furnaces and rolling mills. Closing down the coke ovens would be possible with a new electric arc furnace. But the modernization plan did not mandate the ovens to close until July 1, 1988.[20]

Finally, federal attention turned to the need to clean up the tar ponds. The first contract of what was to become a new spin-off industry to Sydney's steel enterprise—engineering and consulting contracts related to cleanup—was let in April 1984. Acres International was hired to provide advice about the scope of the problem and the options for remediation.

Tar Ponds Cleanup Industry Is Born

The extent of the problem was daunting. Preliminary work suggested that the tar ponds contained the equivalent of 540,000 tons dry weight of toxic waste, with between 4.4 and 8.8 million pounds of PAHs. The sludge was estimated to be between 1 and 4 metres deep. A causeway had been constructed along the route of what had once been a ferry passage through the estuary, connecting the steel plant to the other shore at Ferry Street. It had the effect of subdividing the estuary into two bodies of "water," now referred to as the north and south ponds. The causeway still allowed for tidal flow between the two areas and out to sea.

Acres International's report focused almost entirely on the problem created by hundreds of thousands of tons of PAHs. Despite the presence of PCBs in lobster, neither the government nor the consultant thought PCBs were much of a problem.[21]

Unlike the rest of SYSCO's toxic legacy, PCBs*, were not by-products of baking coal or making steel. They were contained in transformer fluid used in the mill. PCBs had become popular as transformer fluid and solvents because of their stability—they did not break down—and because extreme heat left them unchanged. But this very indestructibility would prove to be a nightmare once the dangers of PCBs were recognized. A persistent pollutant, known to be cancer-causing, PCBs also bioaccumulate, or increase in concentration as they move along the food chain. Although they had been banned years before, they were routinely dumped into sewers that led to the estuary, or from use at the railway yard, operated by another Crown corporation, Canadian National Railways, on the opposite bank of the creek.

Acres performed no thorough testing in the tar ponds to determine the composition of the sludge; instead, it took only random bore holes throughout the estuary, through which only small amounts of PCBs were in evidence. Based on an assessment that minimized the potential presence of the chemicals, Acres proposed three possible approaches: leave the mess where it is and bury it, dig the mess up and take it somewhere else, or dig it up and incinerate it on-site. The incineration option had the added bonus of creating 1,464 person-years of work.

However, incineration was only an option if PCBs were *not* a significant problem. Acres estimated that the proposed incinerator would destroy 99.99 per cent of the PAHs. PCBs, however,

* Polychlorinated Biphenyls (PCBs): A commercial chemical developed as an insulating fluid, it was used widely in transformers and for other industrial purposes until it was banned. In chemical structure PCBs resemble furans and dioxins, which in turn resemble natural hormones. PCBs bioaccumulate, are persistent, and are even harder to destroy than PAHs.

nearly indestructible at extreme temperatures, would only have produced airborne dioxins and other poisons if incinerated. But secure in the untested assumption that PCBs were not a factor, the province opted for incineration. Any remediation work to clean up the tar ponds was on hold until the next phase of modernization had been completed.

The company's nightmarish generation of toxic waste slowed in this period. No government action shut down the coke ovens, as they were not required to close until the plant was modernized, but production nonetheless stopped in 1983. Having produced coke far above demand for many years in order to keep the workforce employed, SYSCO had a 282,000-ton coke stockpile. Closing the coke ovens was a temporary decision to allow the existing supplies to be used, and SYSCO planned to reopen the ovens in 1985.

A number of alarm bells were beginning to ring within the federal government. Certain unreleased studies showed a significantly higher cancer rate in Sydney than in the rest of Canada. From 1977 to 1980, Sydney residents had 346.6 cancer deaths per 100,000 in the population, whereas the national average was 196.2.[22] The Bureau of Chemical Hazards at Health and Welfare Canada raised concerns about the extent of cancer-causing substances emitted from the coke ovens.

On April 3, 1985, Roy Hickman from Health and Welfare wrote to his counterpart in Environment Canada's Atlantic regional office. He noted that emissions from SYSCO's coking operations, even when operating at only 31 to 52 per cent capacity, were "twice the concentration measured in the air in Hamilton, Ontario." Hickman warned that reopening the coke ovens "could be expected to result in increases in morbidity and mortality in the coke plant workers and probably in the residents of Sydney."[23] Five months later, Hickman wrote again, warning that "there is sufficient evidence that exposure to coke production is carcinogenic to humans giving rise to lung cancer . . . and that some finite increase in the risk of cancer will exist as a result of uncontrolled coke oven emission to the ambient atmosphere in Sydney."[24]

Environment Canada's regional head, Ed Norrena, wrote a

clear and compelling warning to the provincial deputy minister of the environment: "Any resumption of coking at the existing Sydney batteries will place at risk the health of coke oven workers, the health of the people in the adjacent community, as well as the health of the environment. . . . Should we not be effective in influencing the coke ovens situation, I will give serious consideration to advising federal agencies involved of my concerns in funding future phases of SYSCO modernization."[25]

This was the first statement of a clear and present danger to public health and the environment caused by the coke ovens. But in the absence of adequate environmental laws in Canada or the will to enforce the existing laws, Environment Canada's only threat was to future funding. The Nova Scotia government felt justified in ignoring the risks to the environment and the near certainty that more people would die because of its decision. The cabinet's desire to reopen the coke ovens went unchallenged by its senior medical adviser. Nova Scotia's provincial epidemiologist, Dr. Pierre Lavigne, was far more sanguine than federal officials about the risks to the adjacent population of restarting the coke ovens. "Since the hazardous nature of the coke oven emissions primarily results from long-term exposure," he reported, "the political or social benefits of allowing the coke ovens to resume operations may outweigh the health risks in the short-term."[26]

Lavigne's advice seemed to be given in some sort of vacuum. He was apparently unaware that the community and coke oven workers had been exposed to the toxic waste for decades, and had only been spared emissions for the last two years. Reopening the coke ovens after such a brief hiatus would hardly constitute "short-term" exposure. Renewed exposure might well be the nail in the coffin for people living near the oven sites for many years.

Given that the published medical literature had abundant reports of increased cancers in coke oven workers, Lavigne's complacency was odd. While the connection between working in coke ovens and dying of cancer had been noted since the late eighteenth century, unequivocably convincing work had been published in 1971. A Pennsylvania study of 17 different companies showed that coke oven workers had a 2.5 to 10 times greater risk of lung cancer, likely because of exposure to PAHs.[27] The most respected of British epidemiologists, Sir Richard Doll, had published work in 1972 linking coke ovens to many other forms of cancer. Many other studies confirming the health risks of working in coke ovens, both from cancer and cardiovascular diseases, had followed.[28]

Steel mills in the United States and the rest of Canada took these studies seriously. While no union is likely to applaud Dofasco, for instance, for the pace of its reforms, the Hamilton steel mill at least brought in significant changes to reduce worker exposure to PAHs. By the 1970s, it had reduced coke oven emissions through pollution control. Workers were given respirators to keep PAH gases out of their lungs. The company also provided clean rest areas, kept free of PAH-laden smoke. And Dofasco encouraged worker education about the dangers of coke gases. Meanwhile, Nova Scotia was planning to reopen a state-run coking operation that had no pollution control and no protective equipment for workers except ineffective paper dust masks.[29]

Premier John Buchanan announced the reopening of the coke ovens in November 1985. Buchanan was himself a Sydney native and lawyer, who as a young man had worked as an office boy at the coal company. Teresa Boyd, arguably Sydney's first environmentalist, recalled sending young John on errands when she was working at the Dominion Coal Company. A staunch Liberal, she would tease the affable Conservative premier, even as she lit up "Talkback," the city's popular open-line radio program, to rail against the pollution everyone seemed ready to accept.[30] As for John Buchanan, he saw the issue as straightforward. He dismissed the health statistics as "nothing new." No argument for keeping the coke ovens shut could compete with "a loss of jobs at SYSCO and . . . a loss of coal sales for DEVCO."[31]

While DEVCO had to keep up sales, the federal bureaucrat charged with overseeing the eventual cleanup noted that reopening the coke ovens "would increase PAH discharges from Muggah Creek to 3,000 to 4,000 kilograms per year, obliterating the effect of the proposed cleanup." [32] The federal environment minister, Tom McMillan, engaged in a public debate through the media with Premier Buchanan. McMillan complained that starting a cleanup while the coke ovens were still operating was like "digging a hole and filling it at the same time."[33]

Buchanan continued to insist that the coke oven workers' jobs were imperative, even though, by this time, there were fewer than 50 men on the ovens site. In the end, although the premier appeared to buckle under the pressure and agree to shut down the ovens to allow federal funds for cleanup to flow, the coke ovens continued baking coal until 1988.

On November 7, 1986, Tom McMillan, the Nova Scotia minister of environment, Guy Leblanc, and the minister of development, Rollie Thornhill, gathered at a Sydney hotel with an audience of local environmentalists (including Teresa Boyd), steelworkers and other community leaders. A $34.3 million federal-provincial agreement was signed to proceed with the excavation and incineration of the toxic mess in the tar ponds. The coke oven workers were to be guaranteed rights of first refusal for work on the cleanup. The tar ponds were described as the worst toxic waste site in Canada and the second worst in North America.

The cleanup announcement was greeted with enthusiasm. Retired steelworker and city alderman Alex MacInnis had lived for 20 years on the banks of the tar ponds on Intercolonial Street. He told the local paper, "If I had the information I have now, I wouldn't have brought up a family here. . . . This is the only place in the world where you could drop a bomb and cause thousands of dollars of improvement."[34]

A government booklet on the project heralded that "a damaged piece of nature will in time be brought back to life." Government plans encouraged people to think of a future time when the estuary could boast a marina or a little park. "Thanks to a historic 1986 federal-provincial agreement, Sydney, Nova Scotia, will be rid of this environmental blight by the mid 1990s," the booklet promised.[35]

No one could have imagined in 1986 how many things would go wrong and how much money would be wasted before the first cleanup plan would be abandoned.

Oh, the companies come and the companies go,
The way of the world we may never know.
We'll follow the footsteps of those on their way
And still ask for the right to leave or stay.
—Kenzie MacNeil, "The Island"

CHAPTER FIVE
Murphy's Career

Now that the tar ponds cleanup was underway, local environmentalists were supposed to be happy. But with incineration the technology of choice, there were grave concerns.

Incineration posed a direct threat to the neighbouring community. Incomplete combustion of plastics in municipal incinerators was known to produce toxic contaminants. But this incinerator was built to destroy highly toxic PAHs. Any incomplete combustion or, worse, a major accident could distribute cancer-causing substances such as dioxins and furans over the community. What is more, the type of incinerator chosen would be inadequate to destroy PCBs. If PCBs were present in the tar pond sludge, then the incinerator might pose even more danger to the neighbours than the ponds themselves.

Though there was a risk of distributing dioxins, furans and other by-products of incomplete combustion over the neighbourhood, the project was billed as an environmental "cleanup." No one wanted to spend time examining the proposed technology too closely. In any event, the Nova Scotia government continually assured residents that there were no PCBs in the PAH muck.

Representatives of a local Clean Up the Tar Ponds committee had attended the signing of the federal-provincial agreement, including a dedicated environmentalist, Roberta Bruce. As the plans for incineration proceeded, her husband, Bruno Marcocchio, emerged as a persistent thorn in the side of bureaucracy. Working with retired steel workers such as Donnie MacPherson and Ron and Don Deleskie, Marcocchio became convinced that the tar ponds held far more PCBs than the trace amounts estimated by Acres. If the memories of workers could be believed, PCBs were routinely dumped and in large quantities.

Donnie MacPherson recalled the dumping: "For decades, when transformer coolant [PCBs] was changed, as happened periodically, it was simply dumped into the plant's sewer system—if there were only 40–50 gallons of liquid involved—and would empty directly into the tar pond. If much larger quantities were involved, it was poured into a 1000-gallon tank, and was taken from there during the summer to spray on the plant's roads to keep the dust down."[1] The railyards also dumped transformer fluid contaminated with PCBs into the estuary. Workers were unaware that the fluid was dangerous. Unbelievably, they would sometimes take a jar home to rub on their aching joints.

The steelworkers' accounts suggested that the ponds contained masses of PCB materials, not mere trace amounts. Clearly, PCBs were a major problem. Marcocchio demanded that the cleanup project should be the subject of a full environmental assessment with a public hearing process.

Environment Canada, having conducted its own preliminary screening, decided that a full public panel process—optional under an environmental assessment guidelines order—was not required. Acres International was made project manager with the requirement that virtually every component of the project would be sub-contracted to different companies. Work on design and construction was governed by 450 separate contracts.[2] The fluidized bed incinerator was to be built by a consortium created in 1989 called Superburn.

Superburn had never before built such a large incinerator. Nevertheless, Nova Scotia environment minister John Leefe and federal environment minister Lucien Bouchard heralded the choice.[3] A one-of-a-kind dredging device was to be designed and built by Ellicott International of Maryland. SYSCO urged the government to employ as much of the company's workforce as possible, and, in fact, SYSCO received 90 of the cleanup contracts.[4]

As Ian Travers, an Environment Canada official, later commented, Murphy's Law ruled: "Murphy has made a career out of this project. It's been incredible."[5]

The Incinerator That Wouldn't

There was trouble from the beginning. Part of the fire-brick lining for the incinerator was manufactured improperly. Once installed, it had to be replaced. Due to a design error, forced hot-air nozzles at the base of the incinerator's combustion chamber were too small and had to be replaced. Before the nozzle problem had been identified, however, the faulty hot-air blowers created a hot spot on the furnace wall that burned out part of the lining. Design problems with the incinerator's fans burned out bearings and required retrofits of the shafts.

But problems with the incinerator were nothing compared to the fundamentally flawed approach to choosing a site for the incinerator and planning how to deliver sludge to it. These decisions would bring down the whole scheme.

Heat could be used to generate the electricity needed to run the generator. In order to take advantage of surplus heat from available boilers at the steel plant, as well as waste heat from the blast furnaces, the incinerator was built within the steel plant complex. The 30-metre stack for the incinerator was below the neighbouring embankment on which Whitney Pier's elementary school was perched.

Years of dumping slag in the estuary had essentially moved the steel plant inland, and a mile's worth of slag mountains now lay between the incinerator and the estuary. The original plan had been to dig up tar pond material and truck it to the site. Indeed, the incinerator had been designed for that delivery method. However, Environment Canada officials were concerned that trucking the toxic sludge would expose adjoining neighbourhoods to even more airborne contaminants, so the decision was made to bring in a special dredge and pump the ponds.[6] The dredge, a Mud Cat, was mounted on a boat to pump the sludge to a pipeline, which would carry the unevenly textured material a mile, uphill and down, to the incinerator. Acres believed that once the watery sludge, intermingled with solids, was "remolded," it would be "very flowable and pumpable."[7]

If there had ever been a public review of the process, one likes to think a bright high school student would have pointed out that you can't pump sludge a mile up and down hills and around corners. Yet legions of experts, consultants and engineers beavered away trying to retrofit and rejig an impossible system to make it work.

By 1992, the project was already significantly over budget and years behind schedule. The incinerator was supposed to have been fired up and burning toxic sludge by 1990, but tests and repairs were underway. One of the final tests required Acres to sample the tar pond material once again to determine its components.

A Nasty Toxic "Surprise"

In October 1992, a bombshell hit. Acres International's testing revealed a "hot spot" of PCBs with concentrations as high as 9 per cent of the total material in the south tar pond, where dredging was to occur first. The hot spot contained an estimated 4,000 tonnes of PCB-contaminated sludge. Under Canadian law, anything over 50 parts per million (ppm) of PCBs requires special handling and destruction at temperatures of at least 1,200° Celsius; the tar ponds suddenly turned out to have PCB concentrations as high as 633 ppm. The incinerator, designed to destroy PAHs, only burned at 900° Celsius. Acres's report on the contamination noted that the government had no way of knowing who was responsible for the illegal dumping.

The island's political culture rapidly assimilated the PCB news. By summer, the annual musical revue, "Rise and Follies of Cape Breton Island," featured a skit with singer Max MacDonald as a grandfather and singer Raylene Rankin as an earnest granddaughter warning her elderly relative of the threat of PCBs: "PCBs, Poppa! PCBs!!" To which the old man in the rocker responded in his thick Cape Breton brogue: "PCBs . . . Progressive Conservative Bastards!"

If the government was surprised by the discovery of PCBs at high concentrations, local environmentalists were not. They were vindicated after years of demanding thorough testing for the deadly chemical. They renewed that call, noting that the "hot spot" discovery had, once again, been the result of a random sample. As for the government's claim that it was not possible to determine the source of the PCBs, Bruno Marcocchio was incredulous: "They say they can't determine who was responsible for the spill when they find PCBs at the end of a pipe leading to Sydney Steel?"[8]

Not only was SYSCO never charged or fined, but bailouts to the plant continued. It now had a debt of $700 million and the province was getting fed up. Increasingly, the cabinet looked favourably at the idea of selling the Crown corporation. But first more money would be needed to get the steel mill into an attractive condition for prospective buyers. The province assumed SYSCO's $700 million debt, and in March 1993 bought 2,400 hectares of land from the corporation, including "a surplus site and area" on SYSCO property.[9] The "surplus site" turned out to include much of the tar ponds and slag heaps where the estuary used to be.

The opposition parties attacked the government of Premier Donald Cameron for spending $1.6 million to buy land from itself. As Cameron had pledged not to grant any further money to

SYSCO, New Democrat critic John Holm charged the government with dishonest dealings: "This is an underhanded way to provide additional subsidies to SYSCO in such a way that they won't be making [Premier] Don Cameron into a liar."[10]

Meanwhile, the incinerator and the pumping and dredging systems were all having problems. An overall review of the project in August 1994 by yet another consulting company, R.V. Anderson, found that "the system had still not been extensively tested, yet there had been blockages in the pipeline between the transfer station and the incinerators. The pipeline gauges . . . had failed at times; there were not adequate facilities to quickly remove blockages, nor were there sufficient pumping facilities with sufficient water available to flush the pipeline at self-scouring velocities."[11]

The incinerator was capable of burning sludge, but the pipe clogged so quickly that the system was nearly inoperable. "There would be two days' supply of sludge and the incinerator would run out and we'd shift to coal [to keep the system running]. It was gobbling up tons and tons of valuable coal," recalled Gary Campbell, the provincial employee with Transportation and Public Works who oversaw efforts to get the system working. "The line kept plugging. We just couldn't get it going effectively, it was costing a fortune, and producing 6 tons of ash an hour."[12]

Self-Denial and Blame-Shifting

In February 1994, ministerial responsibility for the tar ponds cleanup changed hands. The ministers of finance and economic development passed responsibility for the whole mess to Supply and Services Minister Wayne Adams, Nova Scotia's only Black cabinet member. "No choice," was the minister's response to reporters' questions about how he got saddled with the lead role.[13]

By the fall of 1994, the system was no closer to working. Yet another consultant was brought in to advise. J.C. Giroux reviewed the situation and found faults in the design, construction and operation of the dredge and pumping system. His report noted that "self-denial and blame-shifting is taking the place of real problem-solving."[14]

Yet despite all of the problems, Nova Scotia proceeded to accept ownership of the situation. In 1991, the province had created yet another Crown corporation to which it planned to transfer ownership of the incinerator. (Sydney Tar Ponds Clean-Up Inc., or STPCUI, was soon referred to by local environmentalists simply as "Stink.") Under the terms of the contract with Acres, the incinerator was a "turnkey" project—that is, once the new owner turned the key and took over, the builders could walk away. To ensure that the province, and its Crown corporation, were not buying a pig in a poke, the contract called for a 30-day reliability test. Under contract terms, the province would not have to accept the incinerator and dredge scheme until it could be demonstrated to operate to specifications for a 30-day period.

On September 7, 1994, Supply and Services Minister Wayne Adams held a news conference in Sydney to ceremonially sign the papers transferring ownership of the incinerator system to Sydney Tar Ponds Clean-Up Inc. Reaction ranged from incredulity to rage as questions revealed that the province had volunteered to dispense with the 30-day test. Instead, it accepted ownership based on a 48-hour test, which the system, somehow, limped through. City alderman Jack Pearson commented, "I certainly wouldn't want to buy a turnkey home under those conditions. I'd be waiting a long while to get in." Bruno Marcocchio was more direct, saying that taxpayers had been "kicked in the teeth" by the government when it let Acres and Superburn off the hook for the testing. At the press conference, the minister discounted such concerns. The system would be up and running in a few months, he said.[15]

By the spring, Wayne Adams was wondering if the private sector might come to the rescue, and told the media he had advertised internationally to find someone to help. Meanwhile, the incinerator was not incinerating. The problem of the PCBs remained unsolved, even if the incinerator and its dredge began to function properly, and the whole cleanup plan had cost taxpayers nearly twice the original estimate of $34 million, pushing above $55 million.

The province finally realized that the system could not be made to work. The Department of Justice was asked to look into the potential of a lawsuit to recover some of the $55 million, but to this day no legal action has been taken. Instead, without any public notice, the province let out a tender for bids to clean up the mess. When all the estimates came in at or above $100 million, the province did not review the proposals to select the most feasible, environmentally appropriate choice. Instead, it rejected all proposals outright as too costly. With $55 million down the drain, the Nova Scotia government decided that the maximum amount it was prepared to spend on tar ponds cleanup was $20 million. This arbitrary cap was based not on any reasonable estimate of what it would cost to clean up the toxic soup, but rather on how much money the province thought it could afford. This time, the government avoided a tendering process. It approached a prominent Nova Scotia firm that had bid too high. The province asked the environmental engineering firm of Jacques Whitford what could be done for $20 million or less.

The sole-tender approach with a $20 million ceiling led to Plan B. The new supply and services minister, Gerald O'Malley, did not take the time even to consult John Coady, the mayor of Sydney, before coming to town on January 15, 1996, to unveil the new cleanup plan. The minister briefed the mayor only one hour before his scheduled press conference. Former mayor and current MLA for the area, Manning MacDonald, had to attend the press conference like everybody else to find out what was planned.

Hit-and-Run Press Conference

Gerald O'Malley apparently thought the province's new bargain basement cleanup plan would not be controversial. Either that, or he enjoyed orchestrating events to maximize public outrage. With the tacit admission that $55 million had been wasted on a system that did not work, O'Malley announced the new "encapsulation plan." The mountains of slag next to the tar ponds would be used to fill them in. Add some turf and grass seed, and voilà, a park!

The audience could not contain its anger and disbelief. As reporter Jim Meek recounted in a memorable column, "Gerry O'Malley, his Grecian Formula fading before my eyes, announced here Monday that the province intends to cover up the cleanup at the Sydney Tar Ponds. . . . Bruno Marcocchio called Mr. O'Malley a 'criminal' at some stage of the proceedings, which may explain why the minister of supply and services turned a whiter shade of pale before fleeing the premises."[16]

O'Malley made a quick escape from the near riot without taking a single question. His presentation had been effectively heckled from beginning to end. When he began one sentence, "By the time the approval is granted for the project . . ." a local cynic volunteered, "We'll all be dead." The press conference was lively enough to make national television news that night.

For the first time in years, the community reacted to a government proposal with something very close to one voice. Mayor John Coady was not impressed, telling reporters, "We had a number of concerns as a municipality but, basically, the minister said it's this or nothing." Asked why O'Malley had not consulted local government ahead of time, the mayor answered, "Your guess is as good as mine."[17]

Community meetings and protests sprang up. The Rising Sun 4-H Club made stopping the government's plan a group project. Shirley Christmas started organizing in Membertou, while Bruno Marcocchio blasted the proposal daily on local radio programs.

Meanwhile, the firm of Jacques Whitford, which had been awarded the $20 million contract, teamed up with U.S.-based International Technologies to handle the project. The first phase, for $5 million, required a thorough sampling of the tar ponds to determine how much sludge was contaminated with PCBs—the "characterization and delineation phase." It is beyond belief that no thorough sampling had taken place before 1996—ten years after the initial federal-provincial announcement of the cleanup. According to federal law, PCBs could not simply be buried; they had to be located and treated differently. The plan was now to identify them, scoop them out and truck them to a disposal facility in Quebec. In January 1996, the estimated level of PCBs in the tar ponds was under 5,000 tons.

Through the spring of 1996, press reports gave periodic updates on the new estimates of PCB contamination. By April, the estimate was 15,000 tons. By May, it was 25,000. Even the contractor, Jacques Whitford, began to express concerns about the burial plan. When the estimate of PCBs hit 45,000 tons, the Nova Scotia government immediately ordered a halt to the testing. Now spending its own money, the province decided that the toxic chemicals were likely "federal PCBs." About 71 per cent of the contaminated land was federally owned, belonging to the federal Transportation department as part of the Canadian National railbed. Nova Scotia stopped the testing until the federal government agreed to pay for the tests.[18]

The federal government was trying to stay well away from the tar ponds tar baby, having ended its involvement in the cleanup in the late 1980s. Federally, no one wanted to get involved now, except for a backbench Liberal member of parliament who kept asking for help. Russell MacLellan pushed for some kind of federal role. In particular, he asked Sheila Copps, Canada's environment minister, for a federal environmental assessment.

There is little doubt that open scrutiny would have improved the disastrous plan that wasted $55 million in the last round. Now the "bury it in slag" plan was being treated as a *fait accompli* with no public review. An environmental assessment within the new federal law might be the only way to stop the scheme.

On January 25, 1996, Prime Minister Jean Chrétien shuffled the cabinet, shifting ministers into new responsibilities. Sheila Copps knew she was to be moved out of the environment portfolio and over to heritage. She did not have much time. One of the last letters she signed as minister of the environment was to Nova Scotia's environment minister. In it, she increased federal involvement by signalling that a federal environmental review of the burial plan might be required. A copy of the letter, dated January

24, 1996, was sent to the fax machine at Russell MacLellan's office at 1:25 a.m. on January 25—mere hours before Sheila Copps would change portfolios.[19]

Bureaucrats, both federal and provincial, tried damage control after the minister's letter was sent. There were attempts to downplay it as just one of those "Sheila things," with hints that the letter was PMS-induced. But there was no denying that the letter represented a solid statement of concern from the federal environment minister to a provincial counterpart.

Provincial colleagues pressed the new federal environment minister, Sergio Marchi, to get involved. In April, he told the Halifax media by phone interview that he wasn't very impressed with the "encapsulation option." His own departmental official, Ian Travers, had said publicly that the burial plan had been rejected more than ten years earlier as an inadequate solution. Now Marchi said, "Based on public opinion, I think we have to do better than that proposal."[20]

Marchi's own instinct was to investigate for himself. He had been invited to tour the site by the Rising Sun 4-H. He felt he should accept the invitation, go to Sydney and have a look at the tar ponds. Marchi represented an urban Toronto riding where his Italian ancestry was very important. He was young and ambitious, but devoted to his family life and two young children. New in the portfolio, he reacted with a refreshing common sense approach. If people living near Canada's largest toxic waste site wanted him to see it, he wanted to oblige. No federal environment minister had been to Sydney since Tom McMillan, ten years before.

A Ministerial Visit

The minister told officials that he wanted to visit the tar ponds at some point over the summer of 1996—a simple request that opened a major controversy.

Word spread through the Ottawa system, that vast cadre of senior civil servants who have far more control than the public generally knows. Every department that sensed it had some lingering liability for the Sydney mess—Transport, which owned 71 per cent of the remaining non-slag-filled toxic estuary, Health Canada, Industry, Fisheries—brought in the interdepartmental umpire, the Privy Council Office (PCO). Departmental officials asked the PCO to intervene and ensure that the federal minister of the environment not be allowed to visit the country's largest toxic waste site without a prior authorization from the entire Cabinet.

A high-level meeting was held at PCO with the lone voice of Environment Canada officials making the case that their minister should be allowed to look at the tar ponds. Other departments argued that even a visit would raise expectations of future federal funding to clean up the mess. It was one of the rare occasions when the powerful PCO took Environment Canada's side. Sergio Marchi would be allowed to go to Sydney.

But even with PCO clearance, the visit was not that easy. The lead federal minister for the Atlantic region was also the federal minister of health. David Dingwall was a powerful, connected politician. The environment minister could not be seen to upstage him in his own backyard—even if that backyard was toxic. So David Dingwall had to be lined up to participate as well. However, Sergio Marchi and David Dingwall could hardly tour the tar ponds without engaging their provincial counterparts, who were, after all, fellow Liberals. Overtures were therefore made to the province about a planned visit by two federal ministers.

On August 12, 1996, four ministers came to town: the federal ministers for health and the environment, David Dingwall and Sergio Marchi, and their provincial counterparts, Supply and Services Minister Don Downe, who had become the third minister in less than a year stuck with the file, and Health Minister Bernie Boudreau.

The tour shocked Sergio Marchi. It was a stinking hot summer day, and the site reeked even more than usual because of low tide and the raw sewage outfalls that also dump into Muggah Creek. As local reporter Mary-Ellen MacIntyre described it, "As the sun beat down on the stagnant waters of the tar ponds, the raw sewage hugging the edges of the pond emitted a stomach-turning stench."[21]

Sergio Marchi was appalled. He described the site as "horrific . . . quite scandalous," and added, "No briefing can capture this site."[22] But there were more horrors yet to come. Meeting at a local hotel with residents invited by the two levels of government, Marchi heard first-hand of the high cancer rates. David Muise, a cancer survivor himself, son of steelworker Nelson Muise, represented the Canadian Cancer Society. Muise told Marchi how the out-patients lodge for cancer patients in Halifax is known as "old home week" for Cape Bretoners. Although the facility serves the whole province, Sydney residents are usually the majority population, waiting for surgery and recovering from chemotherapy. At home, for days afterward, Marchi would pace the floor sleeplessly and tell his wife about the human toll in Sydney.[23]

Health Minister Dingwall, in his own neck of the woods, did not express the sense of personal outrage evident in Marchi's comments. In fact, despite years of reports within his own department linking Sydney's high cancer rate to cancer-causing chemicals from the making of steel, Dingwall said that there was no definitive evidence. Within government circles, it was reported that Dingwall had been demanding more and better studies of health problems in Sydney. He committed that his department would continue to study the issue.[24]

Thanks to the sensible request of one person, Sergio Marchi, who thought he should take a look at the mess, the "bury it in the slag" scheme came to an end. The hotel gathering launched the next phase of the tar ponds cleanup industry. N.S. environment minister Don Downe announced that the proposal, while not dead, should be seen as just one option among many. In a cathartic mea culpa, the two levels of government committed never to make decisions in secret again. From now on, the public would play a lead role. Accordingly, the government announced the formation of a new community-government committee, with three levels of government and the people of Sydney working together on a cleanup plan.

Sister Ellen Donovan, representing Sisters for the Earth, a local environmental group of nuns at the meeting, urged that community members pray for success as they faced the monumental job of succeeding where several levels of government had failed. Plan C was launched and so was a new, tortuous process that would be known as the "JAG."

Children still play in the factory town,
The engineer waves on his way down the line.
Bloodlines run deep in the factory town,
Mamma says, "Boy, give it time."

A test of mettle for boys in the open hearth.
A test of mettle, luck and skill.
A test of mettle for the ones left behind.
A test of mettle down at the mill.
—Max MacDonald, "Test of Mettle"

CHAPTER SIX
JAG: Analysis Paralysis

By now, the people of Sydney had witnessed two failed cleanup plans, seen more than $60 million sink below the surface of the ooze in the tar ponds, and watched with growing anxiety as it seemed that everyone in the community was dying of cancer. Little wonder that many viewed the chances of a new, improved approach to cleanup with suspicion.

The new Joint Action Group, the JAG, was formed after the August meeting with Sergio Marchi. The new committee was to be chaired by Dr. Carl (Bucky) Buchanan and would include municipal, provincial and federal government representatives.

Dr. Buchanan's connection to the issue was through his role as chair of both the Cape Breton Wellness Centre and a previous community cancer prevention project funded by the provincial government called Act! for a Healthy Sydney. Act! for a Healthy Sydney was initially proposed in 1991 by Dr. Judy Guernsey of

Dalhousie University, and implemented in 1994. Its goals were to "enhance the quality of life in Sydney through health promotion, positive environmental change and disease prevention." Bucky Buchanan became its chair in April 1996.

The JAG faced tremendous obstacles from the beginning. During the previous ten years, there had been constant tension as health studies piled up. In 1985, Health Canada scientists had conducted a survey of cancer rates in industrial Cape Breton as compared to the rest of the country. The study, led by Drs. Yang Mao, Howard Morrison and Robert Semenciw, covered mortality records from 1971 to 1983, comparing death rates to various causes in a number of industrial Cape Breton communities as reported on death certificates. This meant that a cancer patient who died of pneumonia would not be included. Sydney's mortality rate, which would underestimate the numbers of cancers, was then compared to an "expected" cancer rate. The expected number of cancer deaths was based on the mortality rates in the rest of Nova Scotia.

Unlike earlier work on respiratory illness, the study results were not suppressed as confidential. Sydney residents learned that their own observations had been correct—many more people in their community were dying of cancer and other ills than in the rest of Nova Scotia. While the *Mortality Atlas for Canada* had documented significantly higher cancer rates in Sydney for the period 1973 to 1977, the new study had greater detail and covered a larger amount of data.[1]

The Mao study found "significantly elevated mortality among men ages 35–69 . . . in Sydney for all cancers combined as well as for cancers of the stomach, digestive tract and lung. Among women in Sydney significantly elevated mortality was also found for all cancers combined, as well as specifically for cancers of the stomach, breast and uterus, including cervix."[2]

The significantly higher numbers of premature deaths translated into 25 per cent higher deaths for men and 49 per cent higher for women than the provincial average. Predictably, the study increased public demands for action to stop pollution, to compensate injured workers and to clean up the tar ponds. The provincial government responded almost immediately with its own study.

Damage Control

In 1986, the Nova Scotia government commissioned Dr. Pierre Lavigne to get to the bottom of the increased cancer problem in Sydney. Lavigne was the chief provincial epidemiologist who had advised that the 1985 reopening of the coke ovens was probably justified, as socio-economic benefits would outweigh any health risk.[3] For his new investigation into the cancer epidemic in Sydney, he received very clear instructions: his study could focus on anything that might be causing cancer, *except* pollution.

The earlier Health and Welfare cancer study had suggested the coke ovens and steel mill pollution as a possible cause of elevated mortality in Sydney, just as it had pointed to working in the coal mines for increased death from pneumoconiosis in Glace Bay and New Waterford. Except for noting that cigarette smoking should be investigated as a contributing factor in lung cancer, the Mao study had not hesitated to state probable sources of exposure to carcinogenic substances. The study had opened by stating that it was based on concerns about "possible environmental and occupational health hazards" in industrial Cape Breton.

The Lavigne study reported with unusual candour the extent to which avoiding any mention of toxic waste was politically motivated. The report itself noted that "the former minister of health, Dr. Gerald Sheehy, decided upon an investigation that dealt solely with behavioural risk factors."[4] Attempting to explain the decision to ignore PAHs, arsenic, PCBs and other cancer-causing substances in the Sydney environment, Lavigne noted that "major improvements to the SYSCO plant operations were already being planned . . . a study of environmental factors was somewhat academic. . . . [It] might have taken at least 5 years to complete, and have been very expensive."[5]

The Lavigne study consisted of interviewing over 1,000 people by telephone and having 300 people fill out a nutrition survey. The random sample of people questioned were not asked about health problems or their proximity to the steel plant or coke ovens. Instead, they were asked about salt intake, fatty foods, dietary fibre, alcohol and smoking. Most interviews lasted less than ten minutes.

The study made it official: Cape Bretoners were very bad people. They ate too much salt, consumed too many fatty foods, didn't get enough dietary fibre, drank a bit too much and certainly smoked too much.

The study did not question the fact that residents of Sydney were dying much earlier than their friends in other parts of the province or even in other parts of Cape Breton. Did people in Sydney have such a wildly different "lifestyle" from other Maritimers? Were more chips and gravy consumed in Sydney than in Glace Bay? The answers didn't matter—the government wanted to focus on any possible source but the steel mill. It started preaching the gospel of "lifestyle causes of cancer" to a populace with bad habits.

Meanwhile, Act! for a Healthy Sydney was calling for a cleanup of the coke ovens, a comprehensive soil test of Sydney neighbourhoods and sewage treatment. Union activist Dave Ervin, former president of the Steelworkers' Union, became involved as chair of Act! for a Healthy Sydney, pushing for a focus that went beyond "lifestyle." In the 1980s, Ervin had been part of a committee funded by the United Steelworkers that probed into cancer rates of coke oven workers across Canada. In 1991, along with Donnie MacPherson, Dave began working with Dr. Judy Guernsey from Dalhousie University's Department of Community Health and Epidemiology to advise her on an epidemiological survey to look more closely at cancer in Sydney. The initial step was to compare Sydney cancer rates with other community rates in industrial Cape Breton and with Nova Scotia as a whole.

If cancer was higher in Sydney than in Glace Bay, as the 1985 Health and Welfare study had suggested, then it would be difficult for the government to keep up its pretext that "lifestyle" was the primary culprit.

The political resolve to get the funding for Dr. Guernsey's study came from an unexpected source. Don Deleskie had been demanding compensation for workers who had succumbed to the steel plant's poisons. A former steelworker and briefly a coke oven worker, he had disabling respiratory disease. His mother had died at 37 from cancer, leaving six small children to grow up in the shadow of the coke ovens.

Don had watched with increasing frustration as Sydney residents were told that their bad habits were the problem. He had grown up in Whitney Pier, breathing the unbearably polluted air. His friends were dying, and the Workers' Compensation Board, despite a global medical consensus about the dangers of coke oven emissions, still denied claims for compensation. Don decided to do something dramatic to get the attention of Dr. Ron Stewart, the province's new health minister.

In 1993, Don Deleskie went on a very public hunger strike. By the fourth day, Ron Stewart promised action. He agreed to fund the provincial share of the study proposed by Dr. Guernsey, and it was launched with a promise of matching money from Health Canada—although in the end the federal money never materialized. Dr. Guernsey and community members on the committee wanted the study to reflect local concerns, so as a first step they conducted a broad community survey on health and environment.

But before they could get the population survey of cancer started, political infighting threatened the whole project. Sydney mayor Vince MacLean was on the committee, and he didn't like Dalhousie University's having a lead role. Neither did Dr. Bucky Buchanan, who headed the Cape Breton Wellness Centre at the University College of Cape Breton.

Dr. Buchanan was not a medical doctor—his title reflected a doctorate in physical education. He had shown foresight years earlier by coaching Cape Breton's most important politician, Minister of Health David Dingwall to play hockey. On such connections hang a great deal in Cape Breton politics.

Vince MacLean, in addition to being mayor, former member of the legislature and leader of the Nova Scotia Liberal Party, was on the Cape Breton Hospital Board. The hospital had a new cancer centre to deal with the unusually large number of cancer cases, and

MacLean didn't want any money for cancer leaving the island, whether for treatment or epidemiological research. Both MacLean and Buchanan had good channels to David Dingwall, and Judy Guernsey must have realized that she wasn't being paranoid if she thought a Cape Breton political conspiracy was out to get her.

By the spring of 1996, the project fell apart. Despite the meltdown of Act! for a Healthy Sydney, Dr. Guernsey submitted the epidemiological study proposal to Health Canada that summer. While her proposed methodology for the study was still under review at Health Canada, David Dingwall told local media that Dr. Guernsey's proposal was "flawed." It may be unprecedented for a federal health minister to comment on a proposed study before his department completes its own peer-reviewed assessment.

With her name and professional reputation being dragged through the tar ponds, Judy Guernsey and Dalhousie decided to cut their losses and end their involvement with Act!. Dalhousie offered the remaining two and a half years of funding to UCCB. Dr. Bucky Buchanan was to take over both the funding and Act! for a Healthy Sydney in April 1996.

Within months, Bucky Buchanan would get yet another government appointment and an even larger challenge in community-government shared decision-making. He would be the first chair of the Joint Action Group (JAG). Although the position was labelled "volunteer," Buchanan drew an $80,000 salary while seconded from the University College of Cape Breton. The skills required to break up a brawl on the ice would come in handy in chairing the JAG.

JAGGED Organization

From the start, the JAG was a nearly unworkable structure. Government representatives were full members, participating on committees and able to vote. Anyone from the public could join the round table, and 55 people soon did. The JAG included a round table subset called Steering, and within months numerous working groups were created that quickly developed a language all their own. EDGAR was Environmental Data Gathering and Research. PEP stood for Public Education and Participation. There were committees for health studies, site security, remedial options, planning, governance, human resources, finance and ethics. Dedicated volunteers could spend nearly every night of the week in JAG meetings. Many did. Initially, activist Bruno Marcocchio was elected vice-chair and harmonious co-operation seemed possible.

On January 30, 1997, Sergio Marchi, David Dingwall, and the provincial supply and services minister Don Downe came back to Sydney to announce $1.67 million in funds to run the JAG process. Uncharacteristically for a politician, Marchi stated the obvious reality that others dodged. He told the media that if the toxic disaster had been in his backyard in Toronto, it would not have taken so long to get an effective cleanup.[6] The $1.67 million paid for a fully staffed secretariat, as well as $196,000 for a health study to be conducted by Health Canada, $250,000 for monitoring the sewage problem and for studying contamination in the remaining structures on the coke ovens site, as well as $100,000 for community education.

Many in the audience were sceptical. Don Deleskie, whose hunger strike had moved a previous provincial health minister, asked, "When we walk into the house of someone who is dying of cancer, do we say 'I've got good news for you. We're going to have another cancer study!' Why can't we walk in and say, 'Look, we're sorry for what happened and we can't change that,' but at least we could compensate them, and when they go to their graves, they would know that their families were looked after.'[7]

But compensation was not on the agenda. Neither was relocation of families whose backyards abutted the coke ovens site or the ponds. Only former N.S. health minister Dr. Ron Stewart had ever suggested compensation might be in order. He had said it only once in the media, in 1993, and never again.[8] The only proposal for relocation had come some months earlier from a professor at Dalhousie University in Halifax. Biologist Martin Willison proposed that neighbourhoods in close proximity to the poisons should be moved, as neighbourhoods were in Love Canal, New York.[9] While many residents told the press they would welcome a chance to move, Sydney's mayor, John Coady, threw cold water on the idea, condemning it as "something somebody from Halifax would come up with."[10]

Far from evacuating residents, the government had not taken even the most basic steps to keep people away from the toxic ooze. As Sergio Marchi noted, there was not even a fence to keep people out of the area: "As we were [touring the tar ponds] . . . a number of high school students were walking up and down it. We all had protective boots, and they were there with their running shoes."[11]

It was hard to convince people of the danger when the city had traditionally allowed the visiting amusement park to set up in a ball field alongside the tar ponds. It was a convenient spot, with lots of parking adjacent to one of the malls, which had also been built on top of the poisonous land beside the ponds. When activists in the community raised concerns about bringing children to the nation's largest toxic waste site to take the merry-go-round, the chief medical officer for the province, Dr. Jeff Scott, dismissed their concerns. He would bring his pregnant wife to the tar ponds circus, he said.[12]

Biologist, local activist and JAG member Mark Biagi took his message straight to the schools. He went from classroom to classroom in June of 1997 explaining the risks of breathing PAHs. The children came home and asked their parents to call city hall. Community opposition finally forced the circus to the outskirts of Sydney.

The members of the Joint Action Group were faced with a mammoth task. They started by trying to define the problem. For all the attention on the tar ponds, local residents knew that the problem was larger than the estuary's 700,000 tons of toxic PAH sludge, with its estimated 50,000 ton complement of PCB-laden ooze. The problem included the 60-hectare coke ovens site which drained to the estuary. It also had to include the 60-hectare landfill at the top of the hill, and previous dumps, such as the Marsh dump, now abandoned.

Any and all polluting elements of the Muggah Creek watershed had to be considered. Many dumps were on a height of land above the coke ovens, draining down to the creek. As no controls of any kind had existed at the dumps, toxics from the steel operations had made their way there as well. The 13 sewage pipes dumping 3.5 million gallons of raw sewage into the estuary daily were also part of the problem. All told, the JAG members identified 14 separate areas requiring cleanup within the watershed. The Joint Action Group adopted the concept that their concerns were "from the top of the hill down."

This integrated approach made it clear that the problem was far larger than had been recognized ten years before, when the first tar ponds cleanup commitment was made. The dumps drained to a brook that ran orange through the coke ovens site. The site was saturated to depths of 24 metres with all manner of hazardous materials. The ground had been soaked with benzene and naphthalene. Retired steelworkers recalled TNT spills. A large pile of sulphur still sat open to the elements. Walking among abandoned buildings and coke oven stacks, you could almost be knocked over by the sudden whiff of a chemical hazard.

Excavating the coke ovens site and taking samples was not possible until the location of an estimated 160 kilometres of underground piping could be identified. Officials feared that the pipes might still contain sufficient volatile gases to cause an explosion if a drill hit them during sampling.[13] As it was, provincial government staff reported that fires came spontaneously out of the ground.

Sydney itself was heavily contaminated. The ballpark within the Mi'kmaq reserve at Membertou had been made with fill from the toxic area, as was, incidentally, the ground beneath a housing complex in Halifax. The ground and soils surrounding two Whitney Pier schools turned out to have high concentrations of heavy metals and PAHs—hardly surprising given that the schools had been in the path of toxics blowing across the community from the stacks of the coke ovens and steel plant. Arsenic, cadmium, lead and numerous PAHs were found at levels as high as ten times above acceptable guidelines.[14]

Parents were furious—not to discover that the schoolyard was contaminated, but to learn that the school was temporarily closed. Angry parents lined up with local politicians, who feared a conspiracy to close Whitney Pier's schools permanently. As a community that had been denied so much over the years, the residents' first thought was to protect the few community facilities left. Municipal councillor Jim MacLeod vowed to bring angry parents into the streets. "Nothing is wrong with the Pier. No one is growing two heads or glows over here," he declared.[15] The school reopened, and the JAG was left to define what it could do about contamination within the community.

On top of the legacy of past abuse, the city was inviting new forms of contamination. In order to secure $700,000 in funding, Sydney was considering accepting all the biomedical waste from every hospital in Nova Scotia. It planned to incinerate the material in its old municipal incinerator, located at the top of the hill of the Muggah Creek watershed. Activists who toured the incinerator noted that large amounts of the conventional garbage were not totally burned in the ash pile. Fran Morrison, 4-H leader, told the JAG, "I saw plastic milk bags in the ash, some milk still in them and the name of the dairy still visible on the label. The plastic wasn't even melted after going through the incinerator . . . and they want to burn body parts in there?"

With so much contamination, the Joint Action Group started to work on a range of activities, many of them related to establishing itself: developing by-laws, becoming incorporated, and devising rules of procedure, standards for conflict of interest, mediation procedures and ethics guidelines. Negotiating the Memorandum of Understanding (MOU) between the JAG and three levels of government would consume enormous amounts of time and energy, and ultimately would not be signed until September 1998.

Even defining the boundaries of the problem resurfaced as the MOU was negotiated. Having accepted a mandate dealing with the watershed from the top of the hill down, members were puzzled by definitions in the MOU that restricted the area of concern to the tar ponds and the coke ovens, excluding the dumps, the harbour and any neighbourhoods. Government representatives explained that the mandate referred to the areas to be remediated. But lingering questions remained.

The final MOU was a triumph of governmental non-commitment. It was essentially a legal document expressing good intentions, committing governments to "responding to the identified remediation options, constraints on resources at all levels being recognized." Above the signature line for the prime minister of Canada—who ultimately did not sign—and the premier of Nova Scotia were the following words: "This Memorandum of Understanding does not and is not intended to establish legally binding obligations among the Parties, and is to be construed as...a reflection of the dedication of the Parties to the development of the means of addressing [environmental and health] risks."[16]

With so much effort spent on internal policy questions, and so many blockages to moving ahead on cleanup, the Site Security committee, headed by the energetic owner of a local radiator shop, Mark Ferris, got busy. It decided to reduce the physical hazards by fencing in the site and taking down derelict structures.

Like many in the community, Mark already had a history of campaigning for the environment. One day, after his children had been swimming in the harbour near his home close to Point Edward, just outside of town, his daughter came to him complaining, "Daddy, my legs are really itchy." Thinking nothing of it, he headed to his back door, stopping to pick up a complimentary copy of the newspaper. Until that day, Mark had not read newspapers. Having left school after grade nine to earn his own living, he had never seen the point. Now two random events spurred him to activism.

That day's paper included the headline "New sewer line into harbour"—the harbour in which his daughter had just been swimming. Mark then saw a story announcing plans for a new chemical plant. He became enraged that Sydney children were swimming in the hospital, community and steel plant sewer.

Mark wrote his own petition demanding "no more construction in Cape Breton until sewage treatment is in place." And he started going from door to door. Eventually, enough concerned people had joined him to form a new group called Harbour Rescue. As an identifiable group with an environmental concern, Harbour Rescue had been invited to the August 1996 meeting with Sergio Marchi and David Dingwall. Subsequently, Mark became heavily involved with the JAG.

Good Fences Make Good Neighbours

Fencing in the area was harder than anyone would have believed. Steel mesh fence is hugely expensive, and to completely enclose the contaminated area would take more than 30 kilometres of it. Site Security started with the priority areas—places where pedestrian footpaths were routinely used through dangerous areas.

It may seem strange to non-residents that anyone would choose to walk through a toxic waste area. But the geography of Sydney made it nearly inevitable. Whitney Pier had grown up isolated from the rest of the city, a distinct town unto itself. Although shopping malls and the downtown are not far from the Pier, reaching them by road is a long roundabout walk unless you cut through the coke ovens. Mothers with strollers, young people and old people all used the footpaths, even if they complained about coming home with tarry shoes.

The Site Security group put up fences around the coke ovens. To increase respect for the hazard, they chose a clear warning sign to be posted at periodic intervals. It featured the outline of a family, drawn inside a red circle and crossed with a diagonal red bar, and the words Human Health Hazard. Local residents were startled to find that their homes were a stone's throw, or less, from a toxic "human health hazard." Neighbours along Frederick Street on the northern perimeter of the coke ovens regarded the new fence and signs with misgiving. Some worried about how it would affect property values; others worried about the health of their children. Sometimes the sign appeared on utility poles, without benefit of fencing, such as next to the ball field outside the steel plant gates.

Mark brought a remarkable slide show about building the fence to the JAG round table. It had an absurd Bob Newhart quality: "Here's the fencing after we got it up around one side of the coke ovens . . ." (click, click) "Here's where someone's got it down and rolled up . . ." (click, click) "Here's when about half the fence was stolen . . ." (click, click) "And here's a shot where all the fencing is gone."

About $70,000 worth of fencing had already disappeared, and Site Security had to start over. This time, they added to their budget the cost of a security guard for the perimeter.

The Joint Action Group also started pushing for the removal of derelict buildings on the site and huge tanks full of toxic chemicals, abandoned since 1962 when Domtar had shut down operations. But as time passed, the JAG was no closer to finding a solution for cleanup. The planning horizon kept receding into the distance.

It was not long before tempers flared. Bruno Marcocchio and Bucky Buchanan were as much like oil and water as two men could be. They regarded each other with thinly disguised contempt. As the JAG grew, it became its own bureaucracy. Community members who attended the open meetings had to suffer in silence for hours before being allowed to ask questions or make comments. The internal codes and buzzwords of the process would leave any observer arriving for the first time completely in the dark.

As a result, some people around the table increasingly came because they had an interest in other issues besides health. Steelworkers attended to ensure that the Joint Action Group respected the now defunct agreement that union members would get first right of refusal for cleanup jobs. Employees of Sydney Tar Ponds Clean-Up Inc. (STPCUI) attended in hopes that the JAG might come to agree with them that their $55 million incinerator would work fine if the disastrous dredge and pumping system were replaced. Representatives of the University College of Cape Breton kept an eagle eye open for partnering and funding prospects. Employees of various engineering companies attended in hopes that their firm would be in line for rewards whenever the JAG came to select a technology and a company to implement it.

As it grew, the JAG was no longer simply an adviser to government. It saw itself as an implementing agency, in a position to put out tender bids and approve arrangements between government and engineering firms. Unable to resolve preliminary issues, it placed all proposals for clean-up—innovative or traditional methods, crackpot inventions or state-of-the-art pollution-free approaches—in a file drawer to be examined when the JAG was ready.

This delay suited government. Newly elected premier Russell MacLellan, the former member of parliament who had effectively pushed the issue federally, now appeared in no hurry to have the JAG deliver results—or present the bill that would go with them. The new mayor, David Muise, appeared to feel the same way. Ironically, Muise was the same earnest young man who had represented the Canadian Cancer Society and made such an impact on Sergio Marchi at the initiation of the new citizen forum. But as mayor, health could not be his top concern. He was too busy defending the incineration of biomedical waste and the $700,000 a year it brought to the city coffers.

When pressed about tar ponds cleanup, politicians could now defer to the Joint Action Group. Russell MacLellan seemed to ascribe magical powers to the committee. In reverential tones, he would speak of the JAG's volunteer efforts, excusing himself from any comments that might interfere with a process that had "solved" the problem. The JAG had become an effective buffer between public anger and political accountability. And caught in the middle was a group of dedicated people who were increasingly at each other's throats.

Banning Bruno

Bruno Marcocchio took the brunt of the animosity—fuelled much of it as well. He was angry a lot of the time, and not without reason. For a decade, he had been right about every key issue, and either ignored or reviled for his trouble. When he had demanded environmental assessment of the original ill-fated incinerator scheme, he had been ignored. When he had insisted that the sludge had dangerous levels of PCBs, he had been vilified, ignored and eventually proved right. When he had stated that the province had made an unforgivable mistake in abbreviating the incinerator's reliability test to 48 hours, letting private contractors off the hook, once again he had been right. He had predicted that the incinerator would never work, and had been right. He had cursed the "bury it in slag scheme," and in the end the federal government had supported him.

It was small wonder his patience was wearing thin. As well, and more fundamentally, he had lost his wife to cancer in 1992. Roberta Bruce, the daughter of a coal miner, had been involved in Cape Breton's major environmental battles against the use of toxic chemicals to kill forest pests and hardwood trees competing with fir and spruce destined for the pulp mill. She had been one of the first identifiable tar ponds activists. A woman loved for her energy and laughter, Roberta's death had been a blow to all her friends, but to no one did it cause as much pain as to Bruno and the two small children she left behind. Shouldering single parenthood, Bruno had also taken a leading role on community environmental and health issues. But he had one significant disadvantage: he was from Toronto.

It is hard to be truly accepted in tightly knit Cape Breton communities, where conversation often starts "And what was your father's name?" Nearly everyone can be identified based on his or her father's name, and the names of grandfather and great-grandfather are even more assurance of belonging. But Bruno did not come from Cape Breton. To this disadvantage he added a long pony tail, an occasionally aggressive manner and subversive practices such as growing organic vegetables. Many in the community admired him for the strong stands he took. But if one Sydney activist could be easily marginalized, it was Bruno.

In one of the first excommunications of many, Bruno was expelled from the JAG in April 1997. The trigger was an outburst in which Bruno shouted at Bucky—vice-chair to chair—"This has been nothing but an exercise in manipulation and you lied to us."[17] Bruno was referring to Buchanan's failure to raise the contents of a letter that the JAG had received from the provincial deputy minister of supply and services. The letter suggested that Sydney Tar Ponds Clean-Up Inc., a provincial Crown corporation, would only be disbanded if the Joint Action Group assumed control and responsibility for its assets and liabilities. Arguably, the JAG was being set up to take responsibility for the incinerator and the tar ponds mess.

Bucky Buchanan ordered Bruno Marcocchio out of the meeting. "I'm glad to be thrown out by a criminal like you," was Bruno's parting shot. From then on, the municipal government proceeded to treat Bruno as an outcast. By the winter of 1998, the city—through a three-member sub-committee of council—voted to banish him from all municipal property, relying on the provisions of a little-known statute called the Protection of Property Act. By its terms, Bruno was barred from entering municipal buildings or walking on municipal property, such as the sidewalk in front of his home or the city park—for life.

Years before, Sydney Tar Ponds Clean-Up Inc. had taken similar steps, barring Bruno from its premises. Police had actually responded to calls to remove him from the STPCUI offices, even though the public could only read key reports by viewing them at the Crown corporation's offices, which were on private property.

Mayor David Muise defended this draconian infringement on Bruno's civil liberties, saying that the measures were necessary to protect city staff from harassment. Yet no one at the city had ever brought charges against Bruno Marcocchio for any alleged harassment, nor had he ever been charged with anything, although his capacity for verbal abuse was well-honed. On the night of a significant city council meeting, Bruno decided to attend. As he tried to leave his car in the municipal parking lot, he was tackled by six police officers, loaded into a paddy wagon and held in the municipal jail until the city council meeting was over.[18] As one wit observed, the city forced Bruno to continue breaking the law that banned him from municipal property by forcibly containing him within it.

The impact of the ban was significant. Bruno felt unable to protect his children or to attend any event on public municipal land. He feared that he might lose his children if he were convicted. While to the outside world the situation may have seemed ridiculous, for Bruno the nightmare was real.

Months later, the charges against Bruno reached the court. Without difficulty, the judge dismissed them and ruled the banning illegal. For lack of a high-priced lawyer, Bruno never brought counter-charges against the city for violations of his constitutional rights or for false arrest and unlawful confinement. But he did decide to move to British Columbia.

After Bruno's expulsion, others followed. Mark Biagi, the new JAG vice-chair, became the next target for marginalization. Earlier, he had won a battle with the provincial medical officer of health, Jeff Scott, to move the circus away from the tar ponds. Now he was finding the delays and attitude of government officials increasingly hard to bear.

His growing militancy was earning him enemies. His family started receiving threatening telephone calls at home. Once his 11-year-old son answered the phone and had the nasty shock of hearing threats against his father. When the anonymous thug hung up, Mark's son hit the phone code for automatic redial to reach the last caller. Maintaining his composure he told the harasser that he now knew who the man was and that he'd better leave them alone. From then on, threatening calls came only from pay phones.

Mark was unable to find work as a marine biologist in the community. Finally, the combination of unemployment, personal attacks and frustration with the pace of cleaning up the tar ponds tipped the balance, and in the spring of 1998, Mark and his wife took their young family and moved to British Columbia.

Meanwhile, the JAG's internal ethics process removed a number of members. Those with a conflict of interest, such as employees of Sydney Tar Ponds Clean-Up Inc. (now operating as Sydney

Environmental Resources Ltd.), were expelled. Any and all employees of competing engineering firms were removed. A number of community members, including Fran Morrison, who had been representing the 4-H Club, quit in protest over the expulsions. Meetings were testy.

The only Mi'kmaq member of the JAG, Shirley Christmas, left in total despair over the committee's inability to avoid perpetual infighting. "We are going downhill," Shirley told a local reporter. "When I first started at JAG I came for a reason. It's out there," she said, pointing to the contaminated estuary. "That's why I'm here. I'm not here to discuss their [those employed by JAG] finances. I'm here to talk about getting rid of what's out there. . . . We're not children but we act like it."[19]

Though the Joint Action Group attempted to move forward, it approved decisions that moved the community backwards. The 13 sewage outfalls were a recurring theme in discussions. Those who had actually tried to operate the ill-fated dredge swore that no one would be able to stomach working in the toxic waste which was also a cesspool. Members also theorized that the sewage, as it became septic, actually increased the movement of PAHs to the air. Stopping the flow of 3.5 million gallons of raw sewage a day was an immediate priority.

In the initial Acres International assessment of the problem back in 1984, a recommendation had been made to build a giant collector for all the sewage and to pipe it farther out to the middle of the harbour. Dumping huge quantities of untreated sewage into Sydney Harbour would not strike most observers as an environmental cleanup. But the government rationalized the recommendation, pointing out that the sewage was going to end up in the harbour anyway. At the moment, however, it had to travel through the country's largest toxic waste site to get there.

With government support, the collector pipe became an approved-by-JAG plan. It would cost about $10 million, but no one thought the government would spring for another $120 million to build a treatment plant at the end of the pipe. Even though pumping raw sewage into the harbour violated the federal

Fisheries Act, the Canadian government was prepared to share in the cost of building the pipe as the "lesser of two evils."[20] Environment Canada officials speculated on local radio that it would be unlikely that charges would be laid since the federal government was part of the planning process.

But the JAG's most fateful decision would be known by the bland label "Phase One Site Assessment and Remediation." The first contracts were let to further analyze the tar ponds, and to demolish derelict structures on the coke ovens site. Taking advantage of some cost-recovery opportunities, the Phase One process authorized backhoes to move onto the old coal storage area of the coke ovens site and scrape up the 4.5 metres of surface coal for sale to New Brunswick. The work started in the early spring of 1998, with a standard government sign extending credit for the cost-sharing: "A project of the Government of Canada and the Government of Nova Scotia on behalf of the Joint Action Group."

The JAG would appear responsible for the fateful decision to send heavy equipment onto the northern perimeter of the coke ovens site that bordered a string of homes along Frederick Street.

Here comes the drift ice, ho ho ho
Let's get set up for a little more snow
I just stepped out and I froze my toe
It's spring in Sydney, darling.

—Traditional song

CHAPTER SEVEN

Frederick Street

Why were men in moon suits working on the other side of the chain-link fence from the homes on Frederick Street?

In March 1998, in broad daylight, backhoes rolled onto the old coke ovens site and began to excavate the top 4.5 metres of coal. Up until this time, Frederick Street residents remember, there were sporadic excavation activities in the old coal storage area, but they were always clandestine and took place in the middle of the night. But this time, the Joint Action Group had approved the excavation and sale of the commercially valuable old coal for energy production—without considering that the project might disturb the 160 kilometres of underground piping containing lethal gases and chemicals, or might alarm a community that was learning the full extent of the health hazards posed by the coke ovens themselves.

Marlene Kane, community activist on the JAG, repeatedly asked for assurance that the coal would be tested for contaminants before it was taken to New Brunswick to be burned in a power station. Government officials told JAG members there would be testing. However, they omitted mentioning that the testing was for its value as heat (BTU value) to ensure the coal was saleable as fuel. No one tested it for contamination with PAHs and heavy metals, such as lead or cadmium.

JAG's Public Education and Participation committee (PEP) was charged with writing a flyer for local residents to explain the project. It was to be delivered to all nearby homes. Somehow, someone forgot to deliver the flyers to the closest homes along Frederick Street.

And that is why Juanita McKenzie sat in her car, bewildered to see men in environmental hazard suits, masks and gloves riding backhoes and shifting coal within 30 metres of her home.

A Slow Dawning

Juanita's husband, Rickie, is a quiet-spoken Ojibway from Ontario, an artist who sculpts exquisite animal figures and weaves intricate dreamcatchers. He and Juanita, who grew up in North Sydney, acquired their modest Frederick Street bungalow in the mid 1980s for a very reasonable price and the understanding that they would fix it up themselves.

Juanita and Rickie had never before been particularly worried about the location of their home across from the old coke ovens site. When they moved in, the coke ovens were being closed down and city officials were talking about restoring the area as a park. There were plans to cover it with sod and put in a ball field. So they saw fixing up the house as a labour of love, and set to work rewiring, putting in new floors, renovating the kitchen, installing a new roof and insulation, replacing 33 broken water pipes and building a driveway that used 46 tons of slag and 23 tons of rainbow rock, every ounce of which Rickie hauled himself. The house is painted in soft salmon tones, each room lovingly finished with burgundy wallpaper trim. Rickie even put in a small swimming pool for the kids in the backyard.[1]

But some incidents began to worry them. In 1990, benzene was detected in the drinking water. Many Frederick Street residents were given bottled water to drink, although they still had to bathe in the contaminated water. Nine years later, most of the neighbourhood was still on bottled water.

Juanita complained to the city about the coal dust blowing over from the coke ovens site. The fine layer of dust would cover the floors, revealing each morning the clear tracks of little bare feet and their nighttime wanderings. Despite her complaints, no one at city hall took Juanita's concerns seriously. For people who had lived in Sydney when the coke ovens were operating at full tilt, blackening the neighbourhood, and the steel mill was turning the whole city sky orange, a new resident complaining about a little benzene and coal dust seemed like a nuisance.

Other events undermined the McKenzies' sense of security more dramatically. On an August evening in 1995, the family was shaken by a large explosion. Juanita rushed from the house with her video camera to see what was happening. Outside, a thick black smoke blew into her face. The flashes of light through the smoke looked like lightning. Clearly, something had blown up at the steel plant. It was not until morning that they would learn a substation for the steel plant had exploded, and not until much later that the substation held PCB-filled transformers.

A little after midnight, a steel plant worker, wearing a white hazard suit and mask, came to the door to advise them to keep their doors and windows shut—advice the McKenzies hardly needed as Juanita had raced back home through the acrid black smoke. But by one-thirty in the morning the advice was different: evacuate immediately.

Into the car Rickie and Juanita bundled their children and six-week-old granddaughter, Brianna, as well as two friends of their two children staying over for the night—seven people in all—and headed to Bicentennial Hall, across the overpass in Sydney. Juanita cast a worried glance at the gas gauge and wished they had a full tank. As it was, the car was running low as they pulled out into bumper to bumper traffic creeping along Victoria Road to the overpass. "The smell was so bad," she recalls. "The smoke was so thick, and we had to drive right through it."

Whitney Pier's physical isolation from the rest of Sydney—the precariousness of its sole link by the overpass—had never come home to Juanita with such force as it did that night. Their evacuation route forced them to drive through the worst of the smoke, almost directly over the burning substation. Though the windows were rolled up on this stifling hot night, smoke crept into the car. Inching through the blackness, they watched police officers directing traffic, with flimsy paper masks their only protection against toxic smoke. Of course, no one had told them it was toxic, just as no one had warned the firefighters who were below the overpass trying to put out the fires at the substation.

It seemed to take forever to get to Bicentennial Hall. Rickie had thrown on his security guard uniform when the evacuation order came and, as a result of his reassuring presence, he spent the night helping people in the hall who were having difficulty settling into a night as refugees.

Juanita was most concerned for baby Brianna. After the fire, she vomited black liquid and her bowel movements were black, evidence, Juanita believed, of blood in her system. Within two days of the fire, they were at the hospital. Brianna had a viral infection in her lungs. Juanita was convinced it was because of the black smoke the baby had inhaled, but the doctor pooh-poohed her suggestion. Then Juanita pointed out that the baby's diapers were full of unnatural black stuff. Again, the doctor told her she was wrong. He attributed the blackness to something the baby had eaten, even though a six-week-old baby doesn't eat solid food.

After that, Juanita never went to sleep without knowing there was a full tank of gas in the car in readiness for another evacuation. Sure enough, two years later in August 1997 there was another fire, this time at the railway substation. Once again, the family endured a forced evacuation across the overpass. Some local residents were never moved. Juanita discovered that a 92-year-old woman had been left behind as there was no room for her in the hospital. To this day, the overpass is the only route out of the Pier. As Pier residents often say, "God help us if the coke ovens site ever catches fire!"

Then Rickie, who had never had health problems in the past, started to have trouble breathing, and in January 1998 he had a full-blown heart attack. For the next year and a half, he was in and out of the emergency ward, his heart suddenly operating at only 35 per cent capacity.

Making Connections

Although these incidents had worried Juanita, she had not become an activist. She did not rail against the contamination next door to her in the coke ovens—largely because she had no idea that the site was contaminated to depths of 24 metres with a witch's brew of toxic chemicals and poisonous metals. But after the backhoes began their work in March 1998, she started to feel increasingly ill. When Rickie's roses and lilacs—lilacs he had transplanted from the coke ovens—turned pitch black that spring and then dried up and blew away, she became terrified.

Other people in the neighbourhood were also having health problems. Debbie Ouellette was convinced she had a brain tumour. The pain was intense. Even though the doctor offered nothing but Tylenol, Debbie felt this was no ordinary headache. She needed two extra-strength tablets simply to feel strong enough to take two others before she could even get out of bed. She said she felt "dead."[2]

Debbie was not unfamiliar with illness and pain. She is one of that legion of heroic caregivers who tend to the sick and the old in their homes. Her diagnosis of the approaching death of terminal patients often left families astonished. Debbie was often thanked by grieving relatives, who told her she had given them more understanding than had the doctor, and offered more comfort to their dying relatives than had the priest. Debbie knew the look of death, and she was not one to exaggerate her own symptoms. Still, she was scared.

She kept up her usual routine, getting her three kids off to school, seeing her husband to his work with mentally challenged adults and then making her rounds, doing the cleaning, cooking and tending to the needs of the ill and dying.

One Saturday morning with spring's approach, she put on her rubber gloves and wellington boots, and headed into the little brook behind the house for her annual "spring cleaning." Every year, Debbie came down to the brook to clear away debris—papers, old tin cans, anything that had washed down to the part of the brook that flowed along her backyard. But this year, something very different was in the brook. A bright yellowy-orange ooze was in the water, and as Debbie stood there studying it, she realized that it was seeping out of the railbed in front of her. Quickly she stepped out of the brook, took off her gloves and decided she was not going back in.

A chance conversation with Juanita made her think twice about the diagnosis of migraines. Juanita mentioned that she'd been getting killer headaches. Debbie was awash with relief. "So have I!" she exclaimed.

She began making other connections. Her last pregnancy had been very bad; the baby had turned blue and almost died in childbirth. Now her son suffered from asthma and ear infections—he had had eight ear operations in 1998 alone—and shook his head back and forth, banging the sides of his crib continuously. And then there was that weird mouse her older son had found, if it could be called a mouse. It had enormous ears and huge hind legs.

It seemed that almost every pet dog on the street had died of cancer. These were not roving country dogs, but pampered lap dogs, kept on a leash. Yet nine had died within the year on a street with only 17 families. Even dogs from other parts of the city were being affected. Fred Tighe of North Sydney came home from work one night and found his dog, Barney, who often played in the tar ponds, glowing in the dark. "I came in the kitchen and it was dark and there was this thing—well, I couldn't tell it was Barney—this thing just glowing in the corner, kind of yellow like, a big lump glowing," he said.[3]

Other neighbours began to share their stories. Debbie and Ronnie McDonald across the street had bad headaches too. Ronnie was taking medication for blood pressure and said that he thought he was on the verge of a stroke. Unlike the Ouellettes and McKenzies who had moved to Frederick Street after the coke ovens had closed, Debbie McDonald had lived there all her life. She remembers borrowing her father's light blue car in high school and returning it so covered in ore-dust that it was totally black. Her father, Clarence Keller, was one of the workers badly burned and permanently hospitalized in the 1977 coke oven explosion. Ronnie's father had been injured in a coal mining accident and spent most of his adult years in a wheelchair.[4]

Their first baby, a miscarriage, would have been a Down's syndrome child. Their son, Ronnie, spent most of his first years in hospital. "We almost lost him many times," says Debbie. He was born with asthma, allergies, ear infections—for which he has had numerous operations—and a condition called hypospadias, an abnormality in the male urethra. Ronnie Jr.'s penis had a hole at the top, causing him to urinate in two streams. This condition is very rare except in children who live near toxic sites and was found in children born around the Love Canal. A little boy the same age as Ronnie Jr. and from the same Whitney Pier neighbourhood had 21 holes in his penis.

Another neighbour with problems was Louise Desveaux. Her house, located at the farthest end of the street, was serviced by well water that had become contaminated when the municipality rerouted the creek, and toxic contaminants from the landfill leached into her water. She holds up a vial to show what comes out of her tap: the water is freckled with particles and a dark sludge at the bottom stirs like an ancient sea creature. Soon after her water was contaminated, Louise's dog, Bear, started projectile vomiting and passing liquid, bloody stools. Yet both the municipality and the Ministry of Health sent her letters saying that her water met the Canadian drinking water guidelines. Why didn't she move away? "Because my sole income is a monthly pension of $500," she says. "I would be a bag lady if I lost my house."

Now that they were making connections, Debbie asked Juanita to come over and look at the brook. Could the backhoes have caused this ooze? After all, they were working just on the other side of the railbed. Why did the early spring breeze bring such a bad smell?

The answer came unexpectedly when a friend dropped by for a visit. Barbara Lewis had just lost her husband, Al, to cancer. A long-time coke oven worker, Al had been the man sent once to swim through leaking benzene to close valves. Nevertheless, the Workers' Compensation Board (WCB) denied that his cancer was work-related. The WCB still maintained that the coke ovens site had not been a hazardous work place, and had denied benefits to hundreds of dead and dying workers and their families. Barbara was pursuing the claim, even after Al's death. She had found published studies from around the world linking cancer deaths to working in coke ovens.

In the process, Barb had become active for disabled workers against the Workers' Compensation Board, and had testified at hearings of a provincial panel mandated to examine both the Workers' Compensation Act and how it was being administered. She was a fighter, but she wasn't a chemist. Still, she told Juanita and Debbie what they were smelling: "That's benzene." How do you know? they asked her. "Because that's how Al's hair smelled every night when he came home from work," she replied.

Juanita became convinced that the increased odours, the persistent physical symptoms and the sinister ooze in the brook had something to do with the men in environmental hazard suits. She decided to attend an open meeting of the JAG, the citizens' round table, which by now had become the official buffer between elected officials and an angry public. Premier Russell MacLellan was present at the meeting. Juanita approached him without a shred of fear. "My name is Juanita McKenzie and you're going to be hearing a lot more from me, sir!"

Juanita invited JAG program co-ordinator Mike Britten to a meeting at her house on April 30, where he realized for the first time that Frederick Street residents had never been forewarned of the Phase One remediation. While the Joint Action Group was supposed to represent the community, few local residents of Whitney Pier sat on the committee, and no one from the coke ovens area.

Mike Britten had become involved in the JAG process as a volunteer. When the JAG started hiring staff, he left his job teaching engineering at the Coast Guard College. In June 1997, Mike became the JAG's first program co-ordinator. Although he understood the frustration of community members who wanted to see the tar ponds cleanup begin, his hallmark was caution. "The two things that would be most detrimental to the project would be to put a whole chunk of money in it—attracting snake-oil salesmen and driving politicians crazy. All the money could be wasted in make-work projects," he would say. "The other big mistake would be to push ahead with remediation before we know what we are dealing with."[5]

He listened to the residents' complaints, but didn't think they should be too alarmed (although Debbie Ouellette remembers his telling her that it would take 37 million truckloads to remove the contamination from the coke ovens and 50 million loads to remove it from the landfill). The provincial health officer told them to keep their doors and windows closed. Juanita remembers this meeting as the start of a crusade to move their families away from the contaminated street, and contacted the local media to let them know.

Losing patience, Debbie Ouellette also contacted a local Sierra Club member. Before long, the media was tramping through her backyard, looking at the ooze coming out of the railbed and scribbling rapid notes about the neighbours' health problems. Juanita told them, "I'm trying to get out. Would you buy my house? I don't think so. . . . But I just don't want to get out for myself. I want my friends out of here too. I mean, you wouldn't leave a dog on a site like this."[6]

Toxic Testing and Testy Toxic Officials

With press attention, some authorities began to act. Wayne Pierce from the federal environment department went to Frederick Street within days to test the soil in the Ouellettes' backyard and the ooze seeping into the brook.

Environment Canada released the results to the community and the media. The seep behind Debbie's house had very high levels of arsenic. The seep soil had over 222 milligrams per kilogram (mg/kg) of arsenic. The coal ash pile along the street, left behind from either coking or local coal burning, had over 435 mg/kg of arsenic. According to guidelines established by the Canadian Council of Ministers of the Environment (CCME) for residential soil, the Frederick Street readings were more than 18 times above acceptable limits. Backyard soil showed levels of arsenic more than four times higher than the guideline. Environment Canada also found in the seep soil high levels of other toxic heavy metals for which the CCME—a relatively new non-regulatory consultation body between and among governments—had not set any guideline. These included magnesium at 3,396 mg/kg, aluminum at 3,362 mg/kg and manganese at 3,560 mg/kg. The numbers for the coal ash pile were even higher.

The backyards were also tested for PAHs, and not surprisingly, considering all the PAHs dumped during the routine operation of the coke ovens, those numbers too were well above acceptable limits. Backyard soils had naphthalene readings eight times above CCME guidelines; benzopyrene at three times higher than acceptable levels and a host of other aromatic hydrocarbons and PAHs for which no guidelines had been established. Benzene products and naphthalene, like all aromatic hydrocarbon gases, are volatile. They don't merely stay in one place; they move with the air and cause eye irritation and headaches.

While the public expected Environment Canada to take action, both the provincial and federal governments reacted angrily. Nova Scotia did not appreciate the feds taking tests in their province and then releasing the information publicly. Within the federal bureaucracy, Health Canada reacted badly to Environment

Canada's doing anything related to human health. Environment Canada officials had their wrists slapped for having the temerity to determine what was in the yellow ooze.

With the test results in, the Frederick Street neighbours, who had now formed an action group with Juanita as leader, hoped to be relocated. But it wasn't going to be that easy. Whitney Pier MLA Paul MacEwan, a member of the governing Liberals, said that arsenic was not dangerous unless swallowed. "If you ever read any novels about ladies who poison their husbands that way, they put arsenic into their tea. They don't put it on a table in a bowl. You have to swallow it," said MacEwan.[7]

Reporters contacted the Halifax Poison Control Centre, which contradicted MacEwan's reassurances. Arsenic could be dangerous if inhaled, or if absorbed through the skin or the eyes. The nurses at the Poison Control Centre apparently had reference materials other than mystery novels.

Opposition MLA Helen MacDonald of the NDP raised the issue of Frederick Street contamination in the House: "These federal testing reports certainly indicate that this level of concentration is way too high and certainly had nothing to do with natural occurrences. This is definitely unacceptable."[8] MacEwan accused her of political opportunism and "grandstanding."

Debbie Ouellette visited Paul MacEwan at his office and asked directly for his help. His advice to the residents was to go to their insurance companies, ask for estimates on their homes and then have the companies relocate them—because the companies had "top" lawyers who could get the money back from the government. Of course, when Debbie called her insurance company with this suggestion, they laughed at her. She went back to see her MacEwan to tell him that she had voted for him but was losing respect for him. MacEwan responded that he didn't know who voted for him but that the election was over and he had others to worry about. Besides, he said, he had lived in the Pier all his life and didn't have cancer. Debbie wrote in her journal that night, "Paul, you did nothing for us."

Cabinet members ducked the issue and left it to Dr. Jeff Scott, the medical officer for the province, to determine whether Frederick Street was a safe place to live. Scott was the same public health official who had favoured allowing the visiting circus to set up next to the tar ponds. Dr. Scott took precautionary action. He told Frederick Street residents not to go into the brook, not to walk in their yards, and to keep windows and doors closed. Otherwise, he said, living on Frederick Street posed no immediate risk.

Cape Breton Development Corporation (DEVCO), responsible for the railbed, came and dug out the slope where the arsenic ooze had been visible and covered it over with gravel. Provincial government officials came and put a new fence even closer to Frederick Street backyards than the previous chain-link fence with the posted Human Health Hazard warning. Juanita told the press, "Cape Breton has very intelligent fences. They know how to stop toxic chemicals."

Frederick Street residents were sent to the hospital to give hair, blood and urine samples. Juanita remembers that the hospital staff "freaked" at having to touch them and took great pains to wear gloves at all times. "They made us feel like we should be put in a room and quarantined," she recalled. They had to take "his and hers" gallon jugs home to collect every drop of urine for a 24-hour sample. Rickie kept theirs cool on top of an old ice chest in the basement.

By now, Frederick Street residents were growing more organized and getting ever angrier. Ronnie McDonald made Toxic Zone signs, featuring a skull and crossbones, and posted them around the street. They demanded that the coal removal operation be halted and the backhoes removed. But it wasn't until they threatened to chain themselves to the fence—"Rickie, with his bad heart first!" Juanita says—that the Joint Action Group shut the project down in mid June.

Debbie Ouellette and Juanita and Rickie McKenzie started going to JAG meetings as observers, waiting through hours of internal wrangles before being allowed to ask whether anyone was prepared to help them. A handful of volunteers were pitted against not only the country's largest toxic nightmare, but the combined bureaucracies of the municipal, provincial and federal governments—all euphemistically described as their "partners."

One hot July evening at the Steelworkers' Hall, committed citizen activist Mary Ruth MacLellan responded with a motion to call on governments to move the residents who wished to be relocated. While the JAG mandate was argued to exclude Frederick Street and its problems, Eric Brophy reminded the round table that the health impacts of the toxic waste were on the agenda—even if Frederick Street lay outside the formal boundaries of the cleanup area. The comments around the room were overwhelmingly supportive. While the motion would not be binding, it looked as if its passage was a sure thing.

Then Bucky Buchanan intervened from the chair. He had been receiving phone calls at home, he said, from people who claimed to know that the Frederick Street residents were planning to sue the Joint Action Group for its role in the Phase One remediation, which appeared to have triggered the increased levels of contamination. The mysterious callers even seemed to know that members of the JAG, present that evening, were secretly planning to assist Frederick Street members in their anti-JAG legal efforts. He recommended that no vote on the motion be taken until the committee could obtain legal advice about the advisability of voting a motion in support of helping residents. A motion to table followed quickly. Debbie Ouellette was close to tears. No one had asked her whether Buchanan's anonymous informants were right. The discussion had taken place as though she were not even there, and again she went home disappointed.

The burden on the Frederick Street residents who wanted to move out was unbearable. Some homeowners, who thought the others were making a fuss about nothing, berated the McKenzies, McDonalds, Ouellettes and others, blaming them for reducing their property values by making an issue of the toxic contamination. Everyone was also coping with continued health problems, including headaches and eye irritation.

One mother with a very ill two-year-old daughter went public to get help. Larissa Boone was a healthy little girl until she moved onto Frederick Street in early 1998. Within five months, she was plagued with problems: a right eye swollen shut with pus, a recurrent ear infection and fluid in her lungs. Her mother, Tanya, had to attach her to a plastic respirator in order to spray medicine into her lungs. "Nobody came by. Nobody has told us anything," said her mother. "Wouldn't they say something if it wasn't safe?"[9] Because she was renting, Tanya was able to move herself and her daughter several blocks away from Frederick Street later that year and the ailments cleared up almost overnight.

Love Canal North

Juanita decided to map out a street plagued with health problems. She sat down at her kitchen table with a box of coloured markers and a piece of bristol board. Drawing in the scattered houses along Frederick Street, she then colour-coded by disease the health problems of each family. Before long, the street looked like a patchwork. No home had only one colour—or disease. Juanita was struck by the number of cancers, birth defects and respiratory problems along her street. "What blew my mind," she said, "were the browns—Alzheimer cases. Seven on this street alone."

The family that had lived in Debbie's house had all died of cancer. Residents of Juanita's house, past and present, had ailments that included cancers, brain tumours, skin diseases, breast lumps, asthma, heart attacks and strokes. Juanita realized that her perception of her home and her neighbours' homes had changed drastically since the backhoes rolled onto the coke ovens. She saw the houses now and thought of the illnesses and the deaths.

The stress of trying to get the frightened families moved was particularly hard on Juanita. Rickie's heart condition was precarious, and the polluted air was causing him more serious breathing problems. On hot summer nights, they had to sleep with the windows shut. No one had air conditioning and the heat was stifling. Night after night, Juanita would rush Rickie to the hospital for oxygen, sometimes bringing him home again, other times leaving him there for further monitoring. Meanwhile, one of her daughters was having severe kidney infections with blood in her urine. Yet every morning, regardless of how little sleep she had had, Juanita was off early to her job, running a coffee shop to train mentally challenged adults to function in a work environment.

Juanita was also becoming something of a media star. She was interviewed frequently, and was called at home and at work to comment on every new statement made by a government representative. She got used to keeping track of studies and reports. She used her computer to track down information. She was in daily touch with the Sierra Club in Ottawa, and linked up with the original housewife heroine of toxic waste sites, Lois Gibbs of Love Canal. Juanita also had the full-time job of keeping peace within their community group. Tempers flared. Nerves were shot. Malicious rumours spread that residents of Frederick Street were only out for themselves—wanting a free house when there was nothing wrong with the one they had.

Juanita never knew whether to laugh or cry when she looked at their renovated home, at the backyard pool with its patio stones made of bricks left over from the coke oven batteries—bricks that may be contaminated with dioxin. She had loved every inch of her home until she had realized that living in it might kill her husband. Why didn't she just pack up and leave? "Because we are law-abiding, tax-paying, mortgage-paying working people, responsible adults who pay our bills. We'll do this properly," she says.

Juanita, Debbie and the others drew strength from talking to people who had been through fights like theirs before. Activists at Lois Gibbs's Clearing House on Hazardous Waste in Washington, D.C., told them that every issue like this has its stages. Governments will lie and deny. Not all neighbours will believe you. Sometimes you will wonder whether everyone is right and you are losing your mind. Friends will turn on each other. Often marriages will break up. None of it was going to be easy.

Toxic Zoning

As publicity about the contamination of Frederick Street grew,

Dan and Clotilda Yakimchuk found themselves baffled by how anyone could have been allowed to move onto Frederick Street in the last few years. They tried to piece together a history of the struggle for decent housing in the Pier. Both Dan, who had been an alderman at the time, and Clotilda, who had been a community activist confronting slum housing, remembered the series of decisions made about Frederick Street. Dan rooted through his library in the basement and came up with the file: faded photocopies of the Neighbourhood Improvement Program (NIP) and Residential Rehabilitation Assistance Program (RRAP), government funding for urban renewal, old photographs of houses with no running water and people squeezed into unlivable space, and a confirmation of the zoning decision.

In 1974, both Frederick Street and the next parallel street, Tupper, had been permanently zoned as non-residential to provide a buffer zone between the coke ovens and the community. Houses along Frederick Street had been purchased by the city and demolished. The remaining homes were allowed to stand only during the lifetime of the owners who resisted the buyout. No renovations or new construction was allowed. As soon as the remaining stubborn band of original residents died or moved away, their homes were to be expropriated and demolished. No one should have been allowed to build on Tupper or Frederick streets at all.

Obviously, the zoning had been changed. But how had it happened? Dan kept wondering how he could have missed a newspaper notice of the change. He read the *Cape Breton Post* every day, and as a former alderman, he kept an eagle eye out for municipal issues. Dan and Clo invited an old friend to drop by and help them piece together the housing history of the closest streets to the coke ovens in the Pier.

Edie Gorman had lived for the last seven years on Tupper Street. The first house built on the street had been a Habitat for Humanity house built by the international organization that builds homes for and with the poor. Edie remembered that while her home was being built, part of a neat row of new development, someone had asked if she wanted the soil tested for contamination.

According to her recollection, someone in the city government suggested that she didn't really need testing. "I just was so happy to be getting my house, I never thought a thing of it," Edie said with a laugh. She also remembered the name of the lawyer who had handled the purchase for her. Coincidentally, Russell MacLellan, now premier of the province, had handled all the sales along Tupper Street. The current house leader, Manning MacDonald, had been mayor of Sydney at the time of the zoning change.

The Yakimchuks weren't the only ones prodding their memories about the houses along Tupper Street. Dr. Judy Guernsey had also been concerned about the buildings going in so close to the coke ovens. She had not given up the fight to have a proper epidemiological study conducted on the cancer rates in Sydney, and had succeeded in obtaining a multi-year grant from the United States. As she collected and analyzed data, the news of Frederick Street and its toxic ooze was dominating provincial media. Dr. Guernsey recalled visiting the community in 1991. Where the ground had been excavated for the basements of Tupper Street, the soil was black and tarry. Dr. Guernsey had been so struck by the look of the earth that she had taken photographs. It was hard to believe that anyone would be allowed to build on soil that was clearly contaminated with PAH–coal tar.

Dan and Clo kept looking. With the help of a friend, historian Elizabeth Beaton of the Beaton Institute, they found a researcher willing to plow through the microfiche of the *Cape Breton Post* in the period of the zoning change.[10] Sure enough, Dan was right. There had never been an official notification of the change in zoning, although staff had written a story about it. In 1988, zoning had been changed from non-residential (buffer zone) to residential. The next step was to check the zoning change with the files at city hall. But searching the municipal records did not provide any answers. The file holder was there, but the file was missing.

Deborah Nobes, a reporter with the *New Brunswick Telegraph-Journal*, was given the sketchy information that had been pulled together about the zoning changes. Her dogged pursuit of house leader Manning MacDonald gave her the chance to ask him to explain the zoning change for Frederick Street and Tupper Street that had taken place during his tenure as mayor. He denied there had ever been any zoning change, maintaining that Frederick Street had always been residential. "I don't know what you're talking about a buffer zone. I would imagine that if there was any restriction on giving the property away for housing, then we wouldn't have been allowed to do it. They wanted that land for sure. They didn't say no."[11]

However, the record is clear. Both the 1974 and 1982 municipal plans mention the buffer zones. But in 1988, the city ignored or reversed this decision, and the file documenting the change is missing from city records. Frederick Street resident Peggy MacDonald wonders how the municipal government could have offered the land for housing: "Don't get me wrong. The city gave me the land for free. But they must have known what the land was. That's probably why they gave it to us. The city should have known about this before they gave us the land. Or else, they did know and that's why we got it."

The zoning issue was news to most Frederick Street neighbours. The McKenzies and Ouellettes were shocked to discover that just a few years before they bought their homes, the neighbourhood had essentially been condemned, with no renovations or new construction allowed. Suddenly the irregular pattern of their street made sense. The empty lots had once contained houses—houses that had been torn down by the municipal government.

Debbie McDonald had a more personal shock. At the time of the buyout offer to Frederick Street residents, she and Ronnie had not yet moved next door to her mother. When she discussed it with her mother, Debbie learned for the first time that the city government had wanted to move everyone off the street because it was too close to the coke ovens. Her mother had refused the buyout offer, not wanting to leave her home. But the fact that she had never mentioned the buyout to her daughter, nor warned the young couple before they purchased the house next door, strained an already difficult relationship. Debbie McDonald found herself on the other side of the toxic waste issue from her own mother.

No Immediate Risk

From the very beginning, Dr. Jeff Scott had no difficulty concluding that there was no need to relocate Frederick Street families. His standard response to media questions was, "Based on test results so far, we do not believe the residents are at risk."[12]

But the issue was not going away. Frederick Street stayed in the news. Political pressure built to find an "independent" reassurance that the increasingly angry families of Frederick Street were in no danger. Dr. Scott did what government representatives increasingly do in Canada. Rather than rely on the government's own scientists, Nova Scotia used Health Canada funding to retain an outside consulting firm to conduct a "health risk assessment." CANTOX is the largest such group of consultants in Canada, a for-profit business enterprise started by former government scientists who had been downsized in recent years. Previous clients included the Crop Protection Institute, for which CANTOX had prepared a fact sheet minimizing any adverse effects of DDT.

The concept of a health risk assessment requires some explanation. A key idea is that one can live in close proximity to very toxic or radioactive materials, but face no risk if there is no exposure. A computer model is often used to estimate the potential exposure to chemicals. Risk assessment is a relatively new tool. As former head of the U.S. Environmental Protection Agency, Bill Ruckelshaus has said, "Risk assessment data can be like a tortured spy. If you torture it long enough, it will tell you anything you want to hear."[13] That is why other jurisdictions, such as Ontario, require a peer review of any health risk assessment. Nova Scotia does not.

Health Canada also had guidelines for risk assessments, requiring a multidisciplinary approach, full community consultation and other steps bypassed by CANTOX. In essence, the federal department paid for a document that did not meet its own standards.

The report was released on August 12, 1998, by Dr. Scott. He invited all the Frederick Street neighbours to meet with him and Christine Moore from CANTOX. With the media forced to wait in the hallways, they gathered around a table in a crowded room,

COURTESY OF THE BEATON INSTITUTE

Muggah Creek was a beautiful pastoral estuary before the building of the steel plant at the beginning of this century.

COURTESY OF THE BEATON INSTITUTE

The Dominion Iron and Steel Company, shortly after its opening. Financier Henry Melville Whitney proclaimed on the turning of the first sod that "the establishment of these iron works will be the means of introducing Sydney to the length and breadth of the whole world."

COURTESY OF THE BEATON INSTITUTE

A 1908 postcard showing the steel plant on the banks of Muggah Creek. Compare this to the recent aerial view of the creek to see the extent of infilling with slag.

COURTESY OF THE BEATON INSTITUTE

Four hundred coke ovens were constructed, along with four 250-ton capacity blast furnaces and ten open-hearth furnaces. Clouds of toxic coking gas rose from the ovens when the coal was baked.

Steelworkers in 1920. The opening of the steel plant was a magnet for young men in search of work and the population of Sydney grew by 600 per cent in the 20 years after the plant opening.

COURTESY OF THE BEATON INSTITUTE

RAYMOND DOUCETTE PHOTOGRAPHY, COURTESY OF THE BEATON INSTITUTE

Most of the immigrant workers settled in shacks in the area known initially as the Coke Ovens and later as the community of Whitney Pier. Here a blanket of pollution covers Whitney Pier in the 1970s.

OWEN FITZGERALD, COURTESY OF THE BEATON INSTITUTE

By 1986, when the province of Nova Scotia had been running the steel plant for nearly 20 years, the pollution in the estuary had reached staggering proportions.

COURTESY OF SYDNEY ENVIRONMENTAL RESOURCES

The steel plant and the tar ponds, with the harbour beyond, today. Whitney Pier is in the midsection of the photo, on the right hand side.

RON LEVINE PHOTOGRAPHY

The warning sign suggests that the Tar Ponds are in a "cleanup" phase but progress has ground to a halt. Estimates to recover the site have run as high as $1 to 2 billion.

RON LEVINE PHOTOGRAPHY

The coke ovens continued baking coal until 1988 though the Bureau of Chemical Hazards at Health and Welfare Canada raised concerns about the extent of cancer-causing substances emitted from the ovens. Today, only remnants of the coke ovens remain in a heavily polluted area.

WARREN GORDON

Retired steel workers. Front row, left to right: Ray Rykunyk, Clyde Hoban, Nelson Muise, Lorne MacIntyre. Back row, left to right: Gordon Kiely, Dan Yakimchuk, Don Puddicomb, Harry Muldoon. Photographed at the Sydney Steelworkers Pensioners' Club.

WARREN GORDON

Shirley Christmas, noted poet and spokesperson for Mi'kmaq rights. The steel plant was built on traditional Mi'kmaq lands.

WARREN GORDON

Debbie Ouellette (standing) and Juanita McKenzie, at Juanita's computer, struggled to protect their families' health on Frederick Street.

WARREN GORDON

Dan and Clotilda Yakimchuk on their deck: a marriage of two activists who continue to fight for their community.

WARREN GORDON

Eric and Peggy Brophy: a life of suffering and courage.

WARREN GORDON

Tent City: July 1999. Left to right: Clotilda Yakimchuk, Ada Hearn, Peggy Brophy and Eric Brophy. Premier Russell MacLellan's house is in the background.

WARREN GORDON

Tent City: Rear left to right: Eric Brophy, Cindy Steele, Hayley Steele, Lenny Axworthy, Alicia Steele, Eleanor Axworthy, Larry Nixon. Foreground: Ada Hearn and daughter Quinn.

WARREN GORDON

The houses of Frederick Street have mostly been boarded up but toxic waste continues to seep into the neighbourhood.

WARREN GORDON

Children continue to play in the area in spite of signs warning of health hazards.

with the report before them. Dr. Scott sat at one end, flanked by federal and provincial officials. Ronnie McDonald, who was seated right in front of the document, was stunned when his eyes fell on a sentence in the executive summary. He nudged Debbie and pointed it out: "While several residents have reported a variety of health effects (such as ear infections, kidney infections, general malaise, etc.), many of these effects are infections which are bacterial or viral in nature, and are not likely associated with chemical exposures. These ailments are reasonably common in the general population, and the reported incidence on these may well be within the expected range for Sydney this past winter."[14]

Ronnie was too angry to speak and far too angry to stay in the room for the presentation. He pushed away from the table and, catching Debbie Ouellette's approving glance, said, "We're leaving." Juanita, who was seated next to the door, walked out in solidarity, although she had yet to see the report. When she did, it became clear to her that they had all been betrayed.

They were furious. The national news that night showed Juanita, fighting back tears, slamming down the report and saying, "I can't accept this and I won't accept this." Debbie Ouellette wrote in her journal, "Dr. Jeff Scott will be in trouble some day for what he said today, that we are at no health risk at this time." A CTV reporter asked Dr. Scott why the Human Health Hazard signs had not been removed from the fences near Frederick Street, if there was no immediate health risk as he had just reported. Scott told her that it was his understanding the contamination stopped at the fence.

The CANTOX report was the public reassurance Nova Scotia politicians needed. While the media ended up calling it a "health study," it was nothing of the sort. The document was based on whatever data happened to be lying around, including a local study on garden vegetables that had already been thoroughly discredited. The JAG had commissioned the garden vegetable study, but it was so flawed that members of the Health Studies working group later disowned it publicly. The study had been conducted too late in the fall to use any garden produce except root vegetables. The few carrots collected were then treated and stored improperly, making the research almost useless. JAG members joked bitterly about "the four carrot study." Of course, that didn't stop the study's release from garnering headlines that Whitney Pier residents should eat garden vegetables and not worry about contaminated soil.

CANTOX took the random soil, air, vegetable and any other available reports—some not on the same neighbourhood—and fed them into an exposure model based on the consulting group's assumptions. CANTOX looked at the air, water, soil and vegetables and rated the importance of each as a route, or pathway, for the poisonous materials to invade the residents' bodies. For the Frederick Street health risk assessment, CANTOX decided the biggest risk to children was from eating dirt, while for adults, the biggest risk was from vegetables and then soil.[15]

Therefore, on the basis of inadequate and discredited data, CANTOX decided to concentrate the computer modelling exercise on soil and vegetables. Many who had visited Frederick Street had shared in the residents' health problems. Reporters and supporters regularly noted that, after a brief visit, their eyes were tearing and sore, their head throbbing. And none of the visitors had eaten the soil in any backyard.

Although the report often mentioned the weakness of the data base, the conclusions were not cautious. On the contrary, CANTOX issued its assessment summary without equivocation: "No measurable health effects in local residents are predicted to result from long-term exposure to chemicals in the Frederick Street neighbourhood."[16]

For the Frederick Street health risk assessment, CANTOX spent less than three weeks and received $60,000. No one, other than the corporate entity CANTOX, put his or her name on the report. CANTOX claimed that in the three weeks between when Jeff Scott commissioned it to do the work and when the report was released, 17 experts had contributed to the 600 hours it billed. Neither the consulting group nor the provincial government ever revealed the names of the experts. Nor did any of these professionals volunteer to put their name and reputation behind the report's conclusions.

However, CANTOX was very careful to disclaim responsibility for damages to anyone due to the document. The first nearly full page of the report was a long disclaimer: "CANTOX does not have, and does not accept, any responsibility or duty of care whether based in negligence or otherwise, in relation to the use of this report in whole or in part by any third party. . . . CANTOX does not accept responsibility for damages, if any, suffered by any third party as a result of decisions made or actions based on this report. . . ."[17]

It took nearly six months before an independent peer review of the report was concluded. The Sierra Club of Canada commissioned the International Institute of Concern for Public Health to review the document. Industrial hygienist Roger Dixon, with decades of experience in coke oven and steel mill issues, and Dr. Rosalie Bertell, one of Canada's most respected biostatisticians and community health experts, conducted the review.

They were troubled by the report's opening statement that the "Medical Officer of Health had stated on several occasions that he felt that the contamination . . . did not pose an immediate risk to health,"[18] in effect, pre-empting any other conclusion. Both Dr. Bertell and Roger Dixon placed their professional reputations behind the conclusion that they could not support CANTOX's health risk assessment, "based on the meagre sampling data and the uncertainties contained in the methodologies employed."

In response to the critique by Bertell and Dixon, CANTOX released its own statement, maintaining that nothing in the peer review affected CANTOX's original conclusions. Dr. Jeff Scott maintained, "I have full confidence in Christine Moore who was the scientists (*sic*) with responsibility for providing the report. I am very comfortable that CANTOX is a company with international expertise that has done an expert risk assessment on the Frederick Street situation."[19]

At the release of the peer review, Juanita told the press that the residents felt vindicated. But they were no closer to being moved.

It would take months more grief, new seeping toxic sites and arsenic in the basements, not to mention duelling cancer studies, before they would leave their homes—and still not everyone would get out.

'S e Ceap Breatainn tìr mo ghràidh,
Tìr nan craobh, 's nam beanntan àrd;
'S e Ceap Breatainn tìr mo ghràidh
Tìr is àillidh leinn air thalamh.

Cape Breton is the land of my love
Land of trees and high mountains;
Cape Breton is the land of my love
It is the most beautiful land on earth to us.
—Song by Dan Alex MacDonald

CHAPTER EIGHT
Studied to Death

Within two weeks of the CANTOX report, in late August 1998, a new source of contamination was discovered. One hot evening, chatting on a neighbour's porch, Juanita, Debbie and Rickie smelled something new and powerful. It smelled different from the benzene they could now recognize, but, like benzene, it caused a rapid headache. It was coming from the railbed, upstream from where the ooze had been.

They set out, following their noses to find the source of the strong chemical aroma. Just on the other side of the railway track they found it. Like something from a horror movie, the pool of black tarry ooze was boiling lethargically in the heat. When he noticed that it seemed to be spreading, Rickie took a stick and drove it into the ground at the outside edge of the pool. They then hurried away and vowed to report the latest source of contamination first thing in the morning. Rickie had a particularly violent breathing attack that night and had to spend several hours in the emergency ward.

Later, Wayne Pierce, the Environment Canada representative who had always been so helpful, stood with Rickie looking at what was now a spreading black expanse. Wayne tried to reassure him, "Somebody probably just dumped this up here from an old drum," he said.

"Yeah?" queried Rickie. "Where's the drum? How did someone get a drum up here? You can't drive here. Why would anyone dump tar and then take the drum away with them?" Well, Wayne suggested, maybe it was a DEVCO train derailment. Rickie pointed out that the trains never passed by this spot. Wayne took one last guess. "Perhaps someone brought it up in wheelbarrows and then took them away." He thought for a moment and said, "And then again, maybe not."

Wayne went to work again, taking samples. By September 2, 1998, when Rickie took reporters back to the site, the tar had spread almost 2 metres, and the stick he had planted to mark the edge of the toxic pool was now somewhere near the middle.[1] A reporter almost fainted from the smell and had to be led away. The "creature from the black lagoon" made it onto the national radio news that evening, but the Frederick Street activists still didn't know exactly what was in the tarry mess.

Just days later, the results of the testing were released. The pool was nearly pure PAHs. The level of naphthalene, which Canadian Council of Ministers of the Environment guidelines limited to 0.6 miligrams per kilogram was over 9,940 mg/kg. The numbers were also off the charts for benzopyrenes at nearly 5,000 mg/kg, with a CCME recommended limit of 0.7 mg/kg, and a host of other carcinogenic nasties for which no "acceptable" limits had been established.[2]

Once again, the residents demanded to be relocated. The CANTOX risk assessment had not assessed the likelihood of new seeps and new sources of exposure. Likewise, it had focused only on recent contamination, ignoring the decades-worth of pollution that had been dumped on the community. Once again, Jeff Scott had no problem issuing his "no immediate risk" statement, although the government did put up additional fences.

Despair Sets In

By now, the Sydney tar ponds were becoming famous and Frederick Street was being referred to as "Canada's Love Canal." Residents had to endure the added indignity of busloads of tourists gawking and waving at them as if they were freaks in a sideshow. Children from other streets were not allowed to visit anymore, and kids from the infamous Frederick Street were treated like lepers at school, as if they had a communicable disease.

Ronnie McDonald's job was at risk when DEVCO announced closures. He asked the community group, "Is there a God up there to hear our prayers?" Debbie McDonald, who worked at Wal-Mart, said she went through the routine of each day without any feeling. She had lost her pride in being a Canadian and a Nova Scotian, something she had never thought possible.

Debbie Ouellette was constantly near tears. She took weekly photos of her children because she was sure at least one of them would die and she wanted a record of every phase of their growth. She wrote in her journal, "Say something to me that I can say to the kids around me day after day that will help them to understand why I am so upset. The mental anguish of worrying over my children is traumatic. I cannot close my eyes without worrying when it will end. Most parents worry about what their children will do with their lives. I worry whether mine will have a life at all."

In October, she was rushed to hospital with a strange pain in her side. A surgeon removed "something" from her bowel and said he had never seen anything like it before. "A nurse asked me where I lived," Debbie recalls. "I didn't want to say Frederick Street because I was embarrassed, but I did. She said that it was the toxics that were making us sick."

Rickie McKenzie was getting sicker by the day. "Rarely does anyone come up and say, 'How are you feeling?' " he says. "My heart is operating at rock bottom capacity. If people whose hearts are operating at 100 per cent are sick, what should I do? Wear a sign that says, 'Help. My heart is only operating at 35 per cent?'"

Juanita McKenzie was coming under heavy criticism for calling attention to Sydney's dirty little secret. She coped, she says, by running on straight adrenalin: "I feel like I am carrying a heavy burden. I have to make the community aware. If I speak to 200 people and only two believe me, I have saved two people."

In November, the Select Committee of the Nova Scotia legislature on the Workers' Compensation Act tabled its report to the Legislative Assembly. It outlined in great detail a system plagued by delays, denials and despair, and one that had never accepted responsibility for the plight of workers injured or made ill by years of working at the steel plant. The report also contained a series of powerful recommendations that the Liberal government watered down considerably when it introduced amendments to the Act a month later.

Barb Lewis's case was one terrible tale among many. After her husband, Al, who worked at the coke ovens for ten years, swam through benzene to close valves, he started to experience problems with fatigue, vision and breathing. His own doctor diagnosed him with chronic bronchitis from repeated exposure to industrial smoke and dust. A second doctor found a shadow on one lung and confirmed lung disease. Al filed for workers' compensation in 1988 and the board denied his claim without explanation. He appealed and a long paper war began. But Al died in 1997 of a lung tumour and a massive heart attack.

Barb decided to continue his fight and to introduce the issue of chemicals into the case. She undertook extensive research and provided the Workers' Compensation Board with information that the chemicals Al handled every day were 2,800 to 6,000 times above the acceptable limits set out by the Canadian Council of Ministers of the Environment. The board retaliated with evidence given in a British case launched by eight coal miners against the British Coal Company. Barb did a little digging and found out that the British coal miners had won the case, and that the presiding judge had found the physician's testimony to be ill-considered and inaccurate. When Barb informed the board of these facts, it replied that the British case was irrelevant.

In June 1998, Barb finally got her hearing with the board. The WCB expert, Dr. Merv Shaw, testified that Al was not in touch with enough chemicals to make him ill. Dr. Shaw had never met or examined Al Lewis. Barb was denied survivor benefits and awarded Al's compensation only back to 1992—a breathtaking $9.06 per week. The total cheque came to $2,007. When asked why the board refused to give her and Al proper compensation, Barb Lewis has a simple answer: "If the government of Nova Scotia accepted the responsibility for the chemical contamination it allowed these workers to be exposed to, it would open the floodgates and they will never, never allow this to happen."

As always when it comes to Cape Bretoners, there was humour in some of the testimony. Patty Doyle, a coal miner, injured his back so badly from a fall down a coal shaft that he had to have disc fusion in a series of operations. Workers' Compensation sent him back to work, but he was experiencing excruciating chronic pain. So they sent him to a pain specialist:

> She said, I am going to teach you how to deal with your pain. I said, okay, I'll try anything; I am game. Now, how are we going to do this? Well, she said, you have got to let your mind relax and pretend you're at the beach. Close your eyes, she said, pretend you're sitting on a rock and the waves are splashing back and forth. I said okay. After a bit, I stood up and she said, what's wrong? I said I am going to have to go home. She said, why? Well, I said, that last wave just came in and splashed me. I am soaking wet.
>
> I came the next day with a pair of rubber boots and rain gear. I said that I'd be ready for the wave this time, but she had me out picking blueberries. So I said, again, I am going to have to go home, and she said, why? Well, I said, the woods are too hot for me with all this rain gear on; I am going to have to go home and change clothes. So she gives me a tape to play at home.
>
> In the meantime, what is not so funny about all this is my compensation is cut off. I still wear a neck brace. I wear a TENS unit. I wear a back brace

and I use a cane to catch myself when my back gives out but they cut me off compensation. But, hey, every time I listen to this tape, my pain disappears.[3]

Cancer Hot Spot

Meanwhile, the health studies continued to pile up. The JAG Health Studies working group had asked for more and better cancer information back in November 1996. Health Canada commissioned a study conducted by Drs. Pierre Band and Michel Camus, who released their results in September 1998. The study dealt only with deaths attributed to cancer in Nova Scotia. Therefore, anyone who was diagnosed with cancer and had survived was not included in the study, nor was anyone who had left Cape Breton and died of cancer elsewhere. Still, the results were striking.

Cancer of the esophagus, stomach, colon, pancreas, lung, salivary glands, breast and cervix was significantly higher for Sydney residents than for other Nova Scotians, as was multiple myeloma. Taking into account the mortality data from 1951 to 1994—the longest time frame around any study to date—Sydney had 16 per cent excess mortality compared to the rest of Canada. Moreover, the report noted that the cancer rate continued to be on the rise. Band and Camus also looked at non-cancer diseases, and again people in Sydney were dying more than in the rest of Canada. The incidence of Alzheimer's disease, multiple sclerosis, asthma, diabetes in men, mental illnesses and liver disease was nearly off the charts. Also significantly higher were deaths from cardiovascular diseases, hypertension and respiratory diseases. The only exception was for very young children.[4]

In his Ottawa office, New Democrat member of parliament Peter Mancini, a lawyer from Sydney, read the Health Canada study. It was official—cancer was the number one killer in Sydney and it was snatching the residents' lives away from them and their families far earlier than the national or provincial averages. As the terrible statistics about his hometown sank in, Mancini stared at the report near tears. Not knowing why, the thought struck him that if he were to head over to Health Minister Allan Rock's office that moment, he could catch him and talk about the report.

Not even bothering to pick up the phone, he crossed Wellington Street to the Parliament Hill offices of the minister of health. As Mancini reached for the doorknob, the report tucked under his arm, the door opened as Allan Rock escorted his last set of visitors to the exit. Mancini waited a moment and then asked Rock for a private word. The two men sat in Rock's office as the Cape Breton MP expressed his grief and anger over cancer rates in Sydney, the tar ponds, the coke ovens and the fact that people were still living nearly on top of them. Allan Rock had not yet seen the report his department had commissioned. He appeared genuinely upset. Peter Mancini left Rock with his own copy of the report, and with a sense that the minister of health was also deeply troubled by the severity of health problems in Sydney.

Meanwhile, Dr. Judy Guernsey's team had completed its analysis of the municipal cancer rate in Sydney compared with the rate in five other communities in the city's immediate vicinity. Coal-mining communities such as Glace Bay and New Waterford were only a short hop from Sydney. Families shared the same Scottish and eastern European heritage and had drinking, smoking and dietary habits indistinguishable from those of Sydney's residents. Guernsey's study was the first to examine cancer incidence. This meant that for the first time the actual rate of cancer—not the number of cancer deaths alone—was studied. The study also covered a period including more recent data, from 1979 to 1995.

Her data was frightening. Sydney residents were 45 per cent more likely to develop cancer than the residents of the rest of Nova Scotia. From 1989 to 1995, cancer rates in men were 45 per cent higher, but in women they were 47 per cent higher than in the rest of the province. The study covered a wide range of cancers, preventable and not, lifestyle-related and not. The first presentation of the data would be to a medical conference in Saskatoon.

It was front page news in the Nova Scotia provincial daily, the *Chronicle Herald*: "Cancer risk acute in Sydney."[5] Along with the article was an inset of "Cancer facts" from Guernsey's report. From 1989 to 1995, the cancer rates in Sydney, as opposed to in the rest of Nova Scotia, were as follows:

In men:

CANCER OF THE STOMACH	78% higher
COLON AND RECTAL	77% higher
BRAIN	68% higher
PROSTATE	40% higher
BLADDER	39% higher
LUNG	22% higher

In women:

CANCER OF THE STOMACH	78% higher
CERVICAL	134% higher
BRAIN	72% higher
BREAST	57% higher
LUNG	40% higher

The cancer rates in the nearby industrial communities of Glace Bay, Dominion, New Waterford, Sydney Mines and North Sydney, while also higher than the provincial average, were nowhere near as elevated as in Sydney. Women in those towns had a cancer incidence 16 per cent above the provincial average, while men were 14 per cent higher.[6]

Judy Guernsey's comments to the newspapers summed up the extraordinary differences found among adjacent communities. "You find an elevated risk like that for occupational populations," she said, "but it's rare to see this level of impact at a community level."[7] She noted that the increase in stomach cancer was particularly interesting, as the medical literature had already reported a specific link between exposure to PAHs and stomach cancer. "There is the potential that there might be an environmental linkage there," she said. "There seems to be more happening than just lifestyle."[8]

In an interview with the university press, she was more specific. "The five communities are very similar to Sydney socio-economically. They are very poor. There are higher unemployment rates in some cases," she said. "Some people have said that the cancer risk in Sydney is due to lifestyle. We do not support that hypothesis. These data control for that."[9]

Dr. Guernsey felt that the next important step was to perform an epidemiological study on the Sydney Steel workforce. Over 28,000 men had been employed at SYSCO at one time or another. She wanted to find the resources needed to study these men, but she also wanted to study the families exposed through living in the neighbourhood and living with the workers, who brought home toxic substances on their work clothes, or as Barb Lewis remembered so poignantly, in his hair.

Juanita McKenzie and the Frederick Street residents resolved that come hell or high water they would not spend another Christmas in their homes. But as the health studies piled up through the fall, the only government reaction was to ask for more studies.

Premier Russell MacLellan called for more studies in response to the release of Dr. Guernsey's results, which he labelled "alarmist."[10] The newly appointed "cancer czar" for Nova Scotia, Dr. Andrew Padmos, reassured the public. "I don't think there's a reason to panic," said the cancer commissioner. "I think there's a reason to be concerned. I think there's lots of reasons to follow up." In looking for answers, Dr. Padmos suggested that Cape Bretoners were not using early screening programs as much as other Nova Scotians, and once again the focus shifted to lifestyle choices.[11] In order to respond to the studies, Padmos convened a panel of three Ontario experts. It took the review panel the winter months to assess the studies.

Christmas approached, and with a sinking feeling the families realized that they would be spending the holiday on Frederick Street. Juanita and Rickie did their best to put on a little festivity for the children, but Ronnie McDonald had not been feeling well and threw up his Christmas dinner. Debbie Ouellette had a more difficult time. In November, her dog, Queenie, was rushed to the vet with blood in her bowel. The vet told her saving the dog would cost over $500. "We sat the kids down and told them they had a choice—the dog or Christmas. They picked the dog," she says.

At the end of February, gathered together at the Steelworkers Hall, the JAG round table was asked to pass a resolution in favour of relocating residents of Frederick Street. Juanita McKenzie had joined the round table, and Debbie Ouellette continued to attend as an observer. They both pleaded for someone to help. Finally, the JAG members agreed to press forward a resolution calling for governments to relocate Frederick Street residents who wished to be moved by no later than June 1, 1999. As government members abstained, the citizen representatives at the table sent a message of unequivocal support to the beleaguered residents of Frederick Street.

But when spring came, the three levels of government had taken no action whatever to abide by this resolution, insisting it was only committed to coming up with a plan to move the residents by June. Moreover, the issues around Frederick Street exploded in different directions: an aborted cleanup at Sydney Steel made news as workers complained of unsafe working conditions; the large grocery retailer in the Sydney mall, Sobey's, became contaminated; more health studies were released; something odd crept into Debbie Ouellette's basement. And a political scandal broke.

Cleanup, Cover-up

Starting in July 1997, a local Cape Breton contractor with high-level contacts in the federal Liberal party, PLI Environmental, started work on what was to be a two-year project. By September 1998, the work halted as the company ran out of money. PLI had received a $7.7 million contract with SYSCO to clean up around derelict areas of the plant. Work at the north and south ends was to include demolition of blast furnace stoves, piers and blast furnaces.[12]

But concerns about the project began almost immediately. One unusual thing was that the workforce of 150 men was to be 100 per cent from the Steelworkers' Union. However, by agreement with the union, the job site was not to operate by union rules. Pay could be lower and, as it turned out, health and safety standards were lax. In what all parties explained was an unrelated move, PLI gave the union a $70,000 donation to fix the union hall roof.

During the summer of 1998, men who had worked for PLI went to the media with their concerns. Donnie Gauthier and others risked being laid off by raising concerns about environmental and health issues at the site. They claimed they had been exposed to asbestos without being provided with proper protective clothing. They also claimed that hazardous materials had been buried without proper liners or any other measures to protect the environment. Finally, they were concerned because their drinking water smelled strange. Originally, it had come from the Lands and Forest provincial government depot, but when the department realized it was not another governmental operation, but a private contractor, it told PLI to pay for the water. PLI chose to find a cheaper source. It turned out the workers were now drinking water from a nearby swamp. PLI admitted that it had not had any health and safety plan.[13]

Local rumours swirled about how PLI could have run out of money and needed a bailout during the course of the project. But the whole mess became more interesting when it spilled over onto the pages of the *Globe and Mail*—complete with a photograph of Prime Minister Jean Chrétien playing golf with Sydney Liberal Louis Friedman.[14]

The story focused on $2 million PLI received from the federal Transitional Jobs Fund, a pot of money designed for sustainable long-term jobs. The two-year cleanup hardly met the criteria. The *Globe* story carried details of accusations in the Nova Scotia legislature. NDP members Frank Corbett and Darrell Dexter explained that a taped telephone conversation between PLI director James Inch and Sydney businessman John Xidos had been turned over to the RCMP. The gist of the discussion was that the $2 million in federal funds had been obtained for PLI through the efforts of the Friedman brothers, Louis and Benny. During the call, Inch spoke of having paid "personal money out" to the

Friedmans to get a positive response from the federal jobs fund. The amount stated in the legislature was $250,000. The relationship between Louis Friedman and Jean Chrétien was highlighted by the June 1995 photo of the smiling duo in the golf cart.

It was not the first time that political connections had been alleged in the allocation of government funds to Cape Breton, and reaction was predictable. Former Sydney mayor and House leader Manning MacDonald responded in the House: "The contract was with the federal government. The contract was with PLI Environmental. If PLI Environmental wanted to pay somebody else some money in that project to help them out with it, it is of no consequence to me."[15] Benny Friedman told CBC Radio: "I don't know anything about this kind of stuff . . . I'm a businessman, I'm not a politician, all this garbage."

By February 2000, the Canadian political scene was obscured by the auditing scandal engulfing the Department of Human Resources. The PLI grant was one of those audited that had yielded "zero" permanent jobs. An *Ottawa Citizen* story updated the ongoing RCMP investigation of influence peddling allegations for which no charges had yet been laid.[16]

The SYSCO cleanup industry had done nothing to boost public confidence. And within days another local controversy erupted.

"Shop Till You Drop"

The Sydney Shopping Mall along Prince Road had the usual array of mall offerings—the bank, Tim Hortons, a pharmacy and Sobey's. Sobey's is the largest grocery store chain in Atlantic Canada and has a lot of political clout. In March 1999, the store at the mall was expanding. While construction crews worked at the rear of the building, pushing back onto the creek that led to the tar ponds, the front section of the store remained open, selling groceries as usual. As the pilings were sunk, tarry muck oozed. The workers became concerned, and the provincial environment department was called in to test. Sure enough, the sludge was PAH-contaminated. The grocery store was being built on toxic waste.

As construction continued, the JAG passed a motion calling for a stop-work order.[17] In any other part of Canada, the discovery of carcinogenic substances on a work site, especially one that was connected to an operating grocery store, would have resulted in a halt in construction. Not in Nova Scotia. Terry MacPherson of the provincial environment department told the media that there were no reasonable or probable grounds to believe irreparable adverse effects were occurring on the construction site. "We don't feel a stop-work order is prudent," said MacPherson.[18]

Meanwhile, the building was going up faster than any other construction in Cape Breton. Sobey's officials announced plans to build a specially engineered slabfloor and a ventilation system to take PAH-contaminated air outside the building.[19] A lone protester stood outside Sobey's waving a placard to passing cars. One day the sign asked, "Who gave the building permit?" The next, his sign read, "Shop till you drop."

Back to Lifestyle

By April 1999, the panel of experts convened by the Nova Scotia government to review the cancer studies, chaired by Dr. Richard Schabas, former medical officer for the Province of Ontario, released its report. Although agreeing that cancer rates were higher in Sydney than in the rest of Nova Scotia, as well as higher than in other parts of Cape Breton, the panel did not agree with Dr. Guernsey's overall conclusions. For one thing, the report noted, heart disease is also higher in Cape Breton—as though evidence of more than one ailment nullified the cancer findings. "Ecological evidence of this kind cannot prove or disprove the hypothesis that local environmental factors are causing cancers in Sydney," the report stated.[20]

The report also criticized the Health Canada study conducted by Drs. Band and Camus. In response, Band and Camus prepared their own critique of the report, which, while highly technical reading, made clear and concrete the failings of the review. The conclusions drawn in the Band and Camus and

Guernsey studies were criticized on an interpretation of the three studies combined. No specific failings were identified in the studies. The panel acknowledged that their ability to review the studies was "severely hampered" by not having a written report.[20] The Guernsey team's work was assessed by the panel based only on the team's slides shown during a ten-minute presentation. Despite offers from both teams of researchers to provide more data, the panel refused to accept any. In fact, despite the fact that the Guernsey study covered a 17-year period, the panel chose only to examine data in all three studies from the last seven years.

Band and Camus characterized the expert panel's process of judging the studies collectively—without noting that one was based on mortality and another on incidence—as "hazardous at best."

When the panel outlined the most likely ways to reduce the cancer rates in Cape Breton, it did not even list cleaning up the toxic mess. Instead, it recommended addressing obesity, smoking and failure to use early screening programs. Dr. Padmos, the provincial cancer commissioner, joined in dismissing any health concerns associated with the country's largest toxic waste site: "There's a lot of reasons why those environmental issues have to be dealt with, and at the moment, the data that's available to us doesn't add cancer to the list."[22]

Nearly two decades of studies repeatedly confirmed high cancer rates in Sydney as well as higher rates for heart disease, Alzheimer's, mental illnesses, respiratory problems, multiple sclerosis, hypertension and liver disease. A new JAG-commissioned study was about to add birth defects to the list. Released in May 1999, the review covered the birth records from the provincial database from 1988 to 1997. The study confirmed that Sydney babies had a higher rate of birth defects and infant mortality than babies in the rest of the province or in adjacent communities in industrial Cape Breton. Out of a total of 3,852 live births and stillbirths in Sydney, there were 41 deaths within the first 28 days. The provincial average suggested the expected rate would be 29.

The level of major congenital abnormalities was also higher. Of approximately 390 births a year, three more babies in Sydney had serious birth defects every year than would be expected. Particularly elevated were the rates of birth defects for "other central nervous system anomalies,"[23] such as anencephaly and spina bifida. Birth defects had been linked to living near toxic waste sites by a large and credible European study. Pregnant women living 3 kilometres from toxic waste ran a significantly increased risk of birth defects—but this development in the medical literature was not mentioned in the provincial media.[24]

Once again, Nova Scotia's chief medical officer issued reassuring statements through the media. "Anytime there are births with a major congenital abnormality, that's concerning to the mother and family," Dr. Jeff Scott observed. "I think you have to put in perspective [that] the majority of children born in Sydney will not have a major congenital abnormality."[25] As a statement of public health philosophy, Scott's remark was stunning. Apparently, no one should become overly concerned about the rate of birth defects in Sydney as long as most babies are born healthy.

Spring Blooms and Toxics Ooze

Residents of Frederick street greeted the spring of 1999 with apprehension. All winter they had dreaded the thaw and the increased level of pollution it would bring. Once again the brook was leaching a toxic yellow ooze. Once again the residents demanded that they not be forced to spend another hot summer in this fetid place. And once again, as if in some scripted ritual, the government told them they were overreacting. MLA Paul MacEwan said that the province was not responsible for their plight, and if they wanted to move, they would have to do so on their own ticket. He added that if the government were to pick up the tab for them, it would be liable to do so for countless others. While living on the site was no doubt a "nuisance," there was still no proof the area was unfit for human habitation, MacEwan insisted.[26]

But the situation took a decided turn for the worse in late April, when Debbie Ouellette discovered a yellow-orange liquid seeping into her basement from three different spots in the foundation.

Alarmed, she invited Juanita over to look at it. Juanita was horrified. She warned Debbie not to go down there and urged her to padlock her basement door. Then Juanita ran back to check her own basement. Sure enough, there was evidence of a now dried yellow substance in the McKenzie basement as well.

Debbie hated the idea of her basement joining the "must see" sites on the toxic tour. To avoid a barrage of reporters trooping through her house, she took a videotape of the ooze on her basement floor. She gave the tape to Juanita, who started phoning reporters to come to her house to view the video. With word spreading, *Chronicle Herald* reporter Tera Camus did not wait until the appointed hour. Her car loaded with her mother and children, she detoured from a planned family outing and drove straight to Frederick Street. Debbie was just leaving the house. Tera begged Debbie to let her in to take some photographs of the floor. Together they gingerly stepped in and around the colourful puddles. More than half the basement floor was covered.

The front page of the next day's newspaper showed a worried Debbie Ouellette crouched on her basement floor, surrounded with yellow ooze. Juanita McKenzie could see that Debbie was getting close to the edge: "Last year, I was terrified," said Juanita to a local newspaper. "This year I'm mad. It's like we're victims of chemical warfare from our own government." Similar ooze was found in another Frederick Street basement as well as in a home on nearby Laurier Street—four houses in all.

Soon government officials were performing the now all-too-familiar ritual of testing Frederick Street for toxic contamination. But this time they were inside the houses instead of in the backyards. Within ten days, the results came back. To no one's surprise, the substance was contaminated with arsenic. The level approached what had been in the brook the previous spring, and although the CCME has no guidelines for toxic substances found in dwellings, the government was now in a real predicament. Jeff Scott rushed to assuage fears: in a press release, he said the concentration of arsenic in the basements did not pose a threat unless there is "direct contact" on an ongoing basis.

The residents went back to the JAG, asking for specific direction to the government to have them all permanently moved by June 1. Once again, government members abstained, but the resolution was strongly endorsed by the rest.

Within days, the newly appointed minister of the environment, Michel Samson, took action. On May 14, he evacuated families on Frederick Street and its extension, Curry's Lane, to lodgings at the Delta Hotel, although he insisted the move was "temporary" until some study of the basements could take place. He was careful not to admit any liability, but said that he was responding to the high level of anxiety in the community.

Juanita wept in Rickie's arms when she heard the news, although she didn't like the sound of the word "temporary." And she assured the government that she would never set foot in her home again, except to remove her belongings. "There is no more Frederick Street." Debbie Ouellette agreed: "Thank you for giving us a place to stay, Michel, but I'm not moving back." Ronnie McDonald said the pattern of temporary relocation was similar to what happened to families at Love Canal. "They're going to move us and then tell us to go back,." he predicted. "I'm paranoid at this point. It sounds too much like Love Canal to me."[27]

Finally convinced that it would have a full-blown scandal on its hands if the families refused to move back into their homes, the province decided to announce a permanent solution. The federal and provincial ministers responsible had already planned a major announcement in Sydney of new funds for the tar ponds project. But the press release had been prepared without any reference to the plight of Frederick Street residents. Federal environment minister Christine Stewart and Health Minister Allan Rock unveiled a federal-provincial plan to spend $62 million on the tar ponds. Not one penny was allocated for relocating the residents or cleaning up the site. Ten million was allocated to build a sewer collector pipe to dump raw sewage in the harbour. Twelve million was set aside for more health studies. Some money went to the upkeep of the Joint Action Group. And the rest of the funding was to be spent on some on-site technology pilot projects.

The surprise announcement came from the provincial environment minister, Michel Samson. On compassionate grounds, with no admission of liability, the province had decided to relocate the residents of Frederick Street and Curry's Lane permanently.

People who lived on adjacent streets in Whitney Pier were distraught. During the question session with the ministers after the announcement, mothers asked how it was possible for their children to be safe one block away from Frederick Street.

Within weeks, over 40 Pier residents posted For Sale signs in their yards, determined to have their health concerns addressed by the government. They also decided to form an investigative committee. Juanita McKenzie warned the government that their relocation would not be the end of the issue: "This is just the beginning of the biggest problem this government has ever had."

We rise again in the faces of our children,
We rise again in the voices of our song,
We rise again like the waves out on the ocean
And then, we rise again!

—Leon Dubinsky, "We Rise Again"

CHAPTER NINE
The Long Hot Toxic Summer

Jubilation was short-lived over the news that residents of Frederick Street and Curry's Lane were to be relocated. As 24 families worked through the offer from the government, they were frightened that they might not be able to afford it. The province offered residents neither a home elsewhere, nor even the replacement value for the home they were leaving. Rather, the beleaguered families of Frederick Street were offered the assessed value of their homes. With values as low as $30,000 and no house assessed higher than $45,000, they could not find a comparable home elsewhere. While real estate prices in Sydney are well below Toronto or Halifax rates, a modest home costs at least $60,000. The government was essentially offering them a down payment.

As their allowed time in a hotel dwindled to days, Debbie and Ronnie McDonald opted for a smaller home and a bigger mortgage in Glace Bay—one that lacked a working garage, which

Ronnie had used constantly at their house on Frederick Street. Juanita and Rickie McKenzie chose a larger mortgage and a small home just beyond the Sydney downtown. Debbie and Richard Ouellette could find nothing for the money they were offered. Owning their home outright, they couldn't imagine where they could move and how they could afford a mortgage. But none of them were as badly off as one of the other Delta Hotel evacuees.

Anne Ross had lived all her life in her family's home on Laurier Street. A 39-year-old single mother, she held a good job at the provincial department of community services. Although like most people in the Whitney Pier neighbourhood, Anne had followed the Frederick Street controversy with interest, she had never been involved in environmental groups, nor had she thought of demanding relocation for herself and her 13-year-old daughter Lindsay. It was not that she lived far from Frederick Street. Her neat white clapboard home was a stone's throw away, two blocks up. From her kitchen window was a clear view of houses on Tupper Street, with Frederick Street just behind and the coke ovens beyond.

The Arsenic Spreads

In early May, Anne Ross had seen the television coverage of Debbie Ouellette's basement and its yellow ooze. She became alarmed. Her basement floor was also covered with what looked like the same substance. When the discoloured liquid had first leaked into her basement more than a year before, she thought it was rusty water. She had never considered that it might be dangerous until she saw Debbie's basement on television, with a commentary that mentioned arsenic. Anne called the provincial department of the environment to come and test her basement. Sampling took place May 3, 1999.

On May 10, at 4:45 p.m., Anne received a call from the provincial minister of the environment, Michel Samson. Samson, a recent lawyer and, at 26, one of the youngest members of cabinet in provincial history, phoned Anne and asked, "How many people are in your family?" When he learned that she and her daughter were the only two, he told her, "We have a room waiting for you at the Delta."

It turned out that Anne Ross's basement had very high levels of arsenic, as high as had been measured in the seep next to Debbie Ouellette's backyard. At 49.9 mg/kg of arsenic, the levels would have been considered unacceptable by CCME guidelines for outside soil. With government guidelines for arsenic in residential soil at 12 mg/kg, the arsenic in Anne's basement was more than four times above the limit. Not surprisingly, no regulatory agency had established "acceptable" levels of arsenic for a basement—homes are not supposed to be contaminated with poison.

Anne and Lindsay went straight to the Delta and stayed there with the Frederick Street refugees. When the buyout offer was made, initially Anne's lawyer told her to go look for another house. He told all the residents he represented—everyone except the Ouellettes who had remained with the first lawyer to offer help to the residents, Black community activist Rocky Jones—to accept the government offer and get busy finding a new home. But as word of Anne Ross's arsenic basement spread, residents of Tupper Street demanded that they should also be included in the buyout. In the media, Anne Ross had agreed that Tupper Street was also very likely contaminated. As she told reporters, "After all, arsenic doesn't jump. If it's on Frederick Street and Laurier Street, it's on Tupper Street as well."

Selective Compassion

One day, Anne's lawyer told her abruptly, "You're out of it." If the residents of Tupper Street had stayed quiet, he told her, she would have been included in the government's "compassion." But with her support, the Tupper Street families were up in arms. Moving Anne Ross would only intensify pressure on the government to move people on Tupper Street. So, instead of a buyout, the government sent a crew to hose down the Ross basement and then retest. With lower arsenic levels, Anne's home was pronounced safe. But almost immediately, the basement reflooded with yellow ooze. Anne fired her lawyer.

Anne and her daughter watched anxiously as some families moved out of the Delta. The four walls of their room seemed to close in as they wondered where they would go and what they would do. Despite their entreaties, the environment officials maintained that the arsenic was from contaminated fill, not from the coke ovens—as if contaminated fill with high levels of arsenic could have nothing to do with the coke ovens. The chief medical officer for the province, the unflappable Dr. Jeff Scott, informed Anne Ross that there should be no health hazard in moving back into her home—so long as she and Lindsay avoided eating or drinking the toxic substance on the basement floor.

No one ordered any health tests for either Anne or her daughter, though they were both suffering from breathing problems and skin rashes. Not surprisingly, Lindsay had started feeling much better as soon as she moved to the Delta, where her respiratory problems and nosebleeds decreased. Anne had already consulted her doctor about a skin rash she had developed in the last 18 months. Her doctor had told her the splotchy skin was a result of a sun allergy, but now that she was learning about the effects of arsenic, she suspected that it was causing the rash. She vowed never to return to her family home.

City councillor and Tupper Street resident Lorne Green told the local media that the relocation of Frederick Street residents had created widespread panic in his community. The father of a young family, well respected in his constituency, he emerged as a spokesperson for the residents, the lone elected official to respond. With neighbours Todd Marsman and Claudelle MacDonald, Green organized a community meeting and greeted a packed Hankard Street hall on the night of June 6. A new and growing committee of concerned residents was forming.

Stalwarts from previous fights, such as Don Deleskie, explained the situation: "The government knew in 1985 that the residents were going to die of cancer. They knew and they never told us. We've all been the air monitors all these years. We've been sucking the stuff day after day into our lungs. Enough is enough. We're talking about our health!"

Anne Ross recounted to the community her recent experience and why she had fired her lawyer. "We have to stick together as a community," she told them, "for the health and safety of our community."

Debbie Ouellette explained that, despite rumours that people on Frederick Street had received fat cash compensation, the reality was far different. "We can't find anywhere we can afford to move to with what we are being offered," she told the gathering. One person brought up the tens of millions of dollars just announced for demonstration tests of new technologies and for health studies. At the mention of more new studies, Laurier Street resident Joe Pettipas summed up the mood of the community: "Study period is over. It's time to write the exam."

Eviction Notice

On June 17, coming home from work, Anne Ross opened her hotel room door and found a letter slipped underneath. It was a brief message from the official at the provincial department of the environment who had taken the samples. He wrote to notify her that as of Saturday, June 19, the province would no longer be responsible for her hotel bill. It was Thursday night. She was to be evicted from the Delta in less than 48 hours.

In the morning, she breakfasted with Debbie Ouellette. The two made plans to hold a meeting by two p.m. in Anne's room. Word was spread to neighbours on Laurier and Tupper streets.

At two p.m., the room was packed. Squeezing in on end tables, beds and along window ledges were city councillor Lorne Green, Dan and Clotilda Yakimchuk and a group of concerned residents. Debate swirled around a number of limited options. Anne was adamant that she would not return to her own home, but by noon the next day she had to have somewhere to go. Lorne's sister Marie Green summed the situation up: "The reality is that you are homeless tomorrow."

Lorne Green suggested that Anne and Lindsay should go to Transition House, a safe house for abused women and their children. "After all," he said, "you are a victim of abuse from the government." The idea had real merit, but Anne rejected it, pointing out that attracting any media attention to a safe house could pose a risk to the women it sheltered.

Dan Yakimchuk suggested that she remain at the Delta, refuse to leave her room and wait until the police come to remove her forcibly. He recalled all the times that union members had held sit-ins in government offices. "Why move? Make them come and force you out," he urged.

While others debated how she could manage to get food or go to work without being locked out as soon as she left the room, Anne rejected this idea as well. "The staff at the Delta have been really good to Lindsay and me," she said. "My quarrel is not with them." Then she launched her own proposal: "I don't see why I shouldn't just leave the Delta and go pitch a tent in front of Russell MacLellan's house."

A buzz started around the room. Russell MacLellan's house was immediately across the street from a huge empty lot where the hospital had once stood. Since the hospital was demolished, the area had become green, had a few trees and was vacant. Sue Mirao and Ada Hearn started to count up who had tents and how fast they could get a tent protest organized. Others still argued that Anne should stay put at the Delta and wait for the police. In the hubbub, the phone rang and Anne answered.

No one noticed at first that Anne was discussing the strategy with someone. As her voice grew firmer, it rose above the debate around her. Everyone heard her say, "My next step is probably, no doubt in my mind, by twelve noon, the tent will be up across the street from the premier's home. . . . It's either there or Transition House. I'm not taking my daughter home, and I'm not living in those conditions."[1] As she replaced the receiver, she told the group that she had been speaking with Tera Camus, the reporter from the *Chronicle Herald.*

Someone commented that she shouldn't have told the press—

now the government would know her plan well in advance. Everyone was aware that the Nova Scotia government had just fallen with the defeat of the minority Liberal party's budget. Anne's looming eviction date was the first Saturday in a surprise summer election campaign. Nova Scotians would go to the polls on July 27.

If Russell MacLellan realized that his government was evicting a single mother and forcing her and her daughter back into an arsenic-contaminated home, he would surely arrange for her to stay in the hotel until after election day. While everyone wanted Anne and her daughter to be removed from their contaminated home, neighbours were worried that a political ploy could leave them in the same spot in which she found herself now—but without the election campaign to generate political pressure to have her moved. Once the word was out through the media, everyone dispersed to find tents, food, protest signs and to spread the word to other residents to join the camp protest.[2]

Tent City

By noon on Saturday, June 19, reporters watched Anne and her neighbours set up camp across the street from the home of the premier of Nova Scotia. Television cameras zoomed in on tent pegs being driven into the earth, lawn chairs set up with beach umbrellas for shade and protest signs placed to face Russell MacLellan's white clapboard house. Doug Clyke struggled with Anne's pup tent, which bore a faded Pepsi-Cola logo. They were setting up the tents on what had been the hospital's asphalt parking lot, so driving in pegs was harder than on the average campsite. One car slowed down at the sight of the protesters. Sue Mirao recognized bagpiper Courtney MacPherson and called out, "Why don't you come and pipe for us?" Before anyone had time to say "highland fling," Courtney was marching in military precision, while squeezing from the ancient instrument of Scottish terror the wailing strains of the "Skye Boat Song" and "Scotland the Brave." Little girls formed a laughing circle of ersatz country dancing, mimicking the steps they had seen in community Celtic gatherings. With tents and banners, the field looked as though a fair was in progress.

Paul Neville arrived bearing coffee. Joe Pettipas's wife, Sandra, known throughout the neighbourhood as Chi-Chi, made two enormous platters of sandwiches. Ada Hearn's mother sent home-made pie. By suppertime, Todd Marsman had delivered free pizzas from his Whitney Pier pizzeria and a party atmosphere unexpectedly prevailed.

Nearly everyone had been nervous about the legality of the camp. As the police car drove up, the chatter died down. Anne strolled over to the city police to ask about her situation. The two officers were genial and friendly. As long as the tent site stayed "dry," without alcohol, and the noise level subsided by eleven p.m. to allow the neighbours to sleep, the officers had no objections to the camp protest.

By twilight, about 30 campers and visitors were at the tent site. While campers were settling in, Premier MacLellan was at the official nomination meeting to confirm his candidacy for member of the legislature representing Cape Breton-North. Media visits had kept the protesters abreast of his impending return to his Hospital Street home. The group planned a respectful intervention, with Anne and two supporters asking the premier for help.

Just before nightfall, a smiling Russell MacLellan emerged from his house, wearing his red campaign golf shirt, and waved to the crowd. He was greeted with muted applause from the protesters. With media circling, Anne Ross asked the premier to get her family out of the poisonous environment. While Russell MacLellan's attitude was sympathetic, he was completely unhelpful. Maintaining that all departmental officials had said arsenic on her floor was not a health risk, he insisted there was nothing he could do. "It's Saturday and I can't reach any of the appropriate ministers before Monday," he said. "Can't you pick up a phone?" asked someone. "Can't you get her back into the Delta?" No, he answered, it was not possible to pick up a phone.With this incredible response, Russell MacLellan gained a new set of next-door neighbours who, except for one night, would stay put for nearly a month.

The discussion that evening, lit by Paul Neville's loaned propane heaters to ward off a cold June night, focused on the premier's performance. Claudelle MacDonald was incredulous. A lifelong resident of Whitney Pier and member of the Black community, she had been exposed to politicians' language as long as she could remember. "I just can't believe it," she exclaimed. "You're never supposed to start an answer by saying, 'No, I can't.' You're supposed to start with things like, 'I understand your concern, I can understand how you feel.' You are never supposed to say, 'I can't.' And that's all he said, over and over, 'I can't, no, I can't. Nothing I can do'!"

At this point, Don Deleskie turned his cane into a microphone and stuck it in front of Doug Clyke. Doug rose to the occasion and impersonated the premier responding to a reporter's question. "No, I am the premier of Nova Scotia, but I am not empowered to use a telephone until I receive notification from the relevant department that a telephone is an option. You can understand my hands are tied. . . . And that's why I can't reach the phone, because of the way my hands are tied. Who tied my hands? This is a question which even I as premier am unable to answer, because until and unless I receive expert advice to the contrary . . ."

Don was laughing so hard that he had to beg Doug to stop. "The tar ponds won't kill me," he gasped. "I'll die from laughing at you." As the hour approached eleven p.m., passing cars continued blowing their horns to support the protest camp. "Quiet," yelled Anne, "you don't want to wake Russell!"

All through Sunday, a steady stream of visitors raised spirits at the campsite. Federal Member of Parliament Michelle Dockrill, representing the New Democrats, dropped by to offer her support. "If it's okay for the premier to have Anne go back to her home," she quipped, "how about he swaps homes with her for a couple of weeks?"[3]

Local media kept a steady watch over Tent City, as it came to be called. Front-page headlines in Monday's papers quoted Anne Ross describing her disappointment at Russell MacLellan's refusal to act. "I was given my eviction notice [from the Delta]," said

Anne. "It is definitely criminal negligence to force me and my daughter back into our home. I'm not a radical or out just to cause trouble—I just want someone to listen, as this is a toxic, a chemical, a poison and it kills."[4]

Within days, there were over a dozen tents, an inflated children's pool—with water brought in by a local supporter—a cooking area with propane stove and ice chests for the frequent donations of food. A guest book was placed on one table where people could add their comments of concern about health. Supporters dropped by and many telephoned one of the cell phones on site.

Ironically, Anne Ross found that she couldn't manage sleeping in a tent and working every day. So she packed her daughter off to relatives in Toronto and went to stay with friends in the Pier. Although she still visited regularly, the camp foundered for about 24 hours after Anne left, but within a day a core of committed women decided to keep it running. Ada Hearn, Sue Mirao and Cindy Steele (who once said with a laugh, "I even get bills addressed to Sydney Steel!"), with the help of their husbands, many neighbours and friends, formed the ever-present core group. Between them, they had about a dozen children and friends of children running around the site. It was appropriate that the camp always rang with the laughter of children because it was love for their children that motivated the protest. They abandoned the comfort and routine of home to raise awareness of the health threats posed by living next to the coke ovens and tar ponds.

Porta-Potty Saga

One of the most difficult comforts of home to do without was the bathroom. Unlike wilderness camping, which has its own protocol for relieving oneself, city camping posed a whole set of different problems. There was literally no place to go. Some of the premier's neighbours offered bathroom facilities when they were home. But the protesters felt awkward knocking on doors after ten p.m. or early in the morning, and driving to local gas stations was a major operation with a pile of children.

The obvious solution was to get a porta-potty. A supporter phoned a rental agency and asked for a porta-potty to be set up at the camp and billed to her credit card. So far, so good, but the rental company had to get permission from someone to set up toilet facilities on the former hospital site. Believing that the tent city was on municipal property, city councillor Lorne Green volunteered to arrange clearance.

To Lorne's surprise, the site belonged not to the city, but to the province, and the province refused to allow a portable toilet on its property. While the protest was not illegal, therefore, providing a bathroom was going to be. Ada Hearn approached the minister of the United church, just down the street, that bordered on the other corner of the large lot. Reverend Doug Pillsworth was immediately sympathetic. But, he explained, he wouldn't want to do anything without checking with the members of his parish. Within a week, Doug Pillsworth confirmed that the protesters were welcome to set up the porta-potty on the edge of the church parking lot, just down the hill from the campsite. Media attention to the goings-on at Tent City was so intense that the porta-potty was front-page news in the *Cape Breton Post*: "Protesters get toilet facilities."[5]

The presence of a porta-potty gave everyone in the camp a sense of victory. The province wanted to make life difficult for them and had not been successful. But even though the portable toilet was a small victory, it was one the province would not allow. Within a week, Doug Pillsworth phoned Ada. He was embarrassed, and reassured her that the church wanted to continue to offer a corner of their parking lot to the camp. The problem was that the church didn't actually own the parking lot land—the province had merely been allowing them the use of provincial land on the old hospital site. It had been made clear to Doug Pillsworth that if the porta-potty stayed, the church could lose its parking privileges.

The protest camp discussed the threat to the church. Not wishing to harm a parish that had been generous and supportive, the group called the rental agency and had the porta-potty removed.

The protesters were back to the old toilet routine. Someone would announce, "I'm making a pee run. Who wants to come?" and then load up the children, spreading the patronage of washrooms evenly around available establishments.

Life at Tent City developed a rhythm of its own. Children enjoyed the family camp-out and older people brought folding chairs to sit for several hours at a time to share their views on the problem. Eric and Peggy Brophy came for long periods every day, as did Don and Ron Deleskie and their sister Sheila. Clotilda and Dan Yakimchuk came as often as they could. The camp created an ideal drop-in centre for concerned residents. Parents marvelled that kids who routinely complained of nothing to do in homes filled with toys, computer games and gadgets never seemed bored in a field with nothing but tents and other kids. As the days went by, the camp developed a sense of permanence. A local mail carrier even promised to deliver any letters addressed to Tent City, Hospital Street.

Trading Health Horror Stories

Conversation around the lanterns at night turned to health problems. Ada, who had grown up on Frederick Street and played basketball in a high school team nicknamed the Coke Ovens Cannonballs, began to wonder why she had never thought it strange to have splitting headaches as a child. She recalled lying on the kitchen floor while her mother applied cold towels, waiting for the pain to subside. She remembered chronic nosebleeds. But she never thought of herself as unhealthy. All the kids in the neighbourhood had headaches and nosebleeds.

Ada's sister had been in a wheelchair all her life with cerebral palsy. Her brother had cancer. And, she wondered, how could any studies of the Sydney population possibly capture all the health problems of the people who had moved away? Ada told the Tent City group about her cousin who had moved to Ontario as a young man. When his twin son and daughter were born, they both had cancer in their eyes. Both left hospital with glass eyes.

Could this tragedy have been caused by his exposure to the toxic waste all his life?

Cindy Steele recalled the premature death of her father, Tom Curran, and of her uncles, and her own young husband's loss of a kidney to cancer—all of them had lived in the shadow of the coke ovens. Cindy was relieved that her husband, John, had given up working at the steel plant and had returned to school, studying environmental science.

Sheila Deleskie, a former teacher, wondered why she seemed to go to more funerals than weddings of her former students. Severe respiratory problems had cut short her own career—she had been forced onto a disability pension at age 36. In those days, with the steel plant belching orange dust and the coke ovens excreting a constant barrage of toxic air and water, it was a miracle anyone could breathe. Still, Sheila recalled that no Sydney doctor had ever told her the air was unhealthy. "The first time I saw a doctor who put two and two together was the specialist in Halifax," she said. "When I told him I lived in Ashby, he said, 'Oh, that's bad. Have you lived there all your life?' And I told him, 'No, I grew up in Whitney Pier,' and he said, 'Oh, that's worse!' "

Anne Ross began to wonder if her daughter's birth defect, her seizures at 18 months and her chronic breathing problems could have been due to the poisons. The one thing Anne knew was that her daughter's breathing problems disappeared when they were living at the Delta Hotel. So had all the health problems of Rickie McKenzie, now safely settled at Howie Centre just outside Sydney. Debbie Ouellette's kids were feeling better too. Her constant fear that her children would die in the night began to recede. But others began to ask questions and ask questions they had never asked aloud before. Mothers related taking children to the doctor because they were having strange nervous twitches while they slept. Local doctors had told more than one parent that such twitching was just a "habit" their child had developed. It was sobering to realize that the state of health in the community had been far below that of other Canadians. When everyone is sick, it is hard to know what "well" is.

By the second week of the protest, Russell MacLellan denounced it as a politically motivated effort orchestrated by the opposition New Democrats. "Really what this is is an NDP sleepover," he said when caught by local reporters at the Sydney airport. "It's to throw as much chaos into the campaign. It's the NDP way of proceeding, to take people's attention away from the real issues."[6]

When asked by reporters about MacLellan's accusation, Ada, unaccustomed to the subtleties of media interviews, answered, "Russell's a pinhead. This is a perfectly clear example of what toxic waste does to somebody, when they think like [the premier]." She went on to add the obvious point about any charge of "orchestrating" the tent protest. "If they were worried about the election, they should have worried about it before they evicted Anne Ross."[7] In her journal, Debbie Ouellette wrote, "Boy, is he going in the wrong direction. What has he done for us? Nothing, nothing, nothing."

Despite the premier's accusation of the NDP, those at the camp did not see their effort as partisan. With the election campaign in full swing, Tent City put out its own questionnaire to the three major political parties. Both Dr. John Hamm, leader of the Progressive Conservatives, and Robert Chisholm, leader of the New Democrats, phoned to express their concern. Neither, however, would visit the camp or commit to the relocation of Anne or other residents. Finally, three days shy of a month-long protest, the stalwart campaigners held a press conference, announcing that this phase of the protest was over, but the fight had only just begun.

Tent City had invigorated a whole group of activists new and old. In addition to committed activists for community health such as the Brophys, Yakimchuks and Deleskies, younger couples such as Ada Hearn and Larry Nixon, John and Cindy Steele, and the Miraos committed themselves to the cause. Some of Tent City's supporters decided to seek election to the JAG round table, for one of the 20 seats reserved for "concerned citizens." Local organizations had their own seats, as did governments. Anne Ross,

Sue Mirao, Debbie Ouellette, Doug Clyke, Mary MacNeil, Lorne Green, Todd Marsman, John Steele and Cindy Steele all decided to run for the round table. All were elected, and for the first time in the JAG's three years of existence, members of the Black community were at the table.

At the annual general meeting on June 26 1999, Bucky Buchanan formally resigned and Dan Fraser, a career military man returning home to Cape Breton for his retirement, was elected chair. The acrimonious atmosphere that pervaded meetings of the Joint Action Group persisted as Mark Ferris accused Dan Fraser of being the "government choice" for chair. Still, there had not been a time in the last three years when as many community members from the coke ovens neighbourhoods were actively engaged in the process.

The Tar Ponds School

Stalwart volunteers like Marlene Kane and Mark Ferris continued their involvement. The protests over the summer had grown to include Marlene's campaign to ensure that the incinerator be removed from the steel plant site. She discovered that under CCME guidelines, incinerators should not be allowed within 1,500 metres of any residential area. Yet, the elementary school was only 600 metres away.[8] The former Don Bosco school had been burnt down before it could be torn down as scheduled—not because it was built in an area of severe contamination, but because a *larger* school was going to be built on the same site. Construction started in the summer of 1999 of a new public-private-partnership—or P3 school—overlooking the tar ponds. Children from the nearby community of South Bar, outside Sydney, would be bused to the new school and the former South Bar school would also be torn down. No one in government seemed to think that building a new school for 670 elementary school children less than 1,000 metres from the country's largest toxic waste site was a problem.

Meanwhile, within days of breaking camp, other protesters who had decided to put their efforts outside the JAG process were in the news again. Ada Hearn had come upon a toxic creek running through the coke ovens in Ashby, a neighbourhood just across the overpass to Sydney.

Through gatherings at Tent City, a number of Ashby residents had connected with activists from Whitney Pier. Ashby was the first neighbourhood on the other side of the overpass connecting the Pier to the rest of Sydney. Just like Frederick Street, Vulcan Avenue in Ashby ran parallel to the old coke ovens site. And just like on Frederick Street, some residents of Vulcan Avenue and other streets in the neighbourhood were becoming increasingly alarmed about the toxic waste next door. A frequent visitor to the camp from Beech Street had breathing problems whenever the wind blew across the coke ovens. And she told Ada that she was particularly worried about the water that flooded their homes after a heavy rain. Ada and Larry set out to investigate the brook behind the Beech Street house. It wasn't hard to find, but it was hard to believe.

The brook ran bright orange. When Ada's sandal-clad feet accidentally became wet, the skin burned off. Over the next few days, Ada and Larry traced the brook upstream to the municipal landfill. From the landfill, it ran south to behind Vulcan Avenue in Ashby, cut across the coke ovens site and flowed along the perimeter of the coke ovens along Frederick Street. When Ada called Environment Canada officials to report the hazard and to ask for a fence around the brook to prevent children from playing there, she was amazed to find she had not "discovered" the problem after all.

The government had tested the brook in 1986 and found high levels of aromatic hydrocarbons and PAHs, including benzopyrene, as well as arsenic, lead, iron and manganese.[9] The JAG had identified the brook as one of the 15 local sites responsible for "ongoing release of contaminants." Yet there were no warning signs in an area still popular for berry picking and all-terrain vehicles. Don Deleskie accompanied Ada and Larry to the brook with reporter Tera Camus from the *Chronicle Herald*. "You think the coke ovens is bad," he mused, "but then you've got the tar ponds, then you got the dump, and then you got this here, a cancer-causing stream, that's wide-open." Councillor Lorne Green was appalled that the stream was not enclosed: "The site is supposed to be secured and the bare fact that you can drive down there in a car amazes me."

On July 27, Nova Scotians went to the polls. Despite all the media attention and protests against the ongoing toxic contamination of Sydney, no political party committed to moving people away from the site. The closest anyone came to taking a tough position on local issues was Dr. John Hamm, leader of the Progressive Conservative Party. Hamm pledged to close down SYSCO, or at least to ensure that no more public money went into the plant. Although pollsters had said the election was too close to call, John Hamm won with an impressive majority of 30 seats, leaving the Liberals and New Democrats in a dead heat for Official Opposition status with 11 seats each. Russell MacLellan held on to his seat, as did the perenially re-elected Liberal Paul MacEwan.

With the election over, Anne Ross was now stuck back in her house, with arsenic pooling on the basement floor. Lindsay was also back from Toronto. Anne decided to have the cracks in the foundation sealed as the government had advised. Of course, the province hadn't offered to pay for repairs, but recommended a contractor at Anne's expense. After he examined her basement, Anne was stunned when the contractor refused to give an estimate until he had done further checking. "I can't send my men in there until I get guarantees from the government that it is not an unsafe workplace. I may need a decontamination unit," he said. The buck was passed from the department of labour to health to the environment, which finally advised that there should no problem working in her basement—as long as the workers wore masks and gloves.

Anne Ross was livid. Grown men couldn't be exposed to arsenic in her basement, but she and her daughter were supposedly in no danger living there? No one in government had arranged for even the most basic health analysis of whether Anne or Lindsay showed signs of exposure to arsenic. It was easier for the government to continue stating that they couldn't have been exposed—as long as they hadn't eaten the arsenic. With the help of the Sierra Club, Anne obtained the protocol for having tests taken of hair and urine for arsenic exposure. As instructed, she took the information to a local doctor. He dismissed her concerns, told her she was wasting his time and suggested that if she wanted a hair sample taken, she should go to her hairdresser.

Anger Builds

Meanwhile, community frustration was mounting. The coke ovens were still leaching poison, the brook still ran orange, and the tar ponds still allowed toxics to waft into the air and migrate daily to the harbour and the ocean beyond. Community meetings had been held. Committees and sub-committees had been formed. And after three years the only planned "cleanup action" was the construction of a collector sewer to divert thousands of gallons of effluent a day from the estuary into the harbour. Even though dumping raw sewage would violate the federal Fisheries Act, Environment Canada officials told local media that they didn't expect Sydney to be charged. After all, three levels of government had approved the collector pipe scheme through the JAG, and the federal and provincial governments had provided $10 million in funding as well.

In early August, a visit by members of the National Round Table on the Environment and Economy (NRTEE), an advisory body appointed by the prime minister, reinforced the need to get the cleanup started. The NRTEE members toured the tar ponds site, the steel mill and its slag heaps, and the coke ovens, then walked up Laurier Street, past the day-care centre, to Anne Ross's house. Anne wasn't home as she was attending a JAG meeting. Lindsay Ross let in the delegation of prominent national appointees, and the group trooped through Anne's immaculate kitchen and chintz-covered living room to the basement door.

Everyone had a chance to stare in disbelief at the yellow ooze on the basement floor.

In a joint meeting later with the JAG, NRTEE members expressed politely their disbelief that three years into the process, the selection of the preferred remediation technology was still a further three years into the future. They urged the JAG to set deadlines accelerating the cleanup. When the discussion turned to the planned dumping of untreated sewage, British Columbia's former premier Mike Harcourt, an NRTEE member, shared with the group his background on the issue. As mayor of Vancouver, he was charged criminally under the Fisheries Act for the city's dumping of untreated sewage. "Criminal charges concentrate the mind wonderfully," he told them. It took a few years to put together the funding required, but the treatment plants were built.

None of the NRTEE members who visited the community and toured the site were prepared for the enormous extent of the contamination or the scale of the challenge. When asked about his impressions, the former head of the Canadian Chemical Producers Association, Jean Belanger, tried to speak of his concern for the community's children. As he began by referring to his own nine grandchildren, he became overcome with emotion and was unable to continue. His tears were shared by many around the room. Mike Harcourt told national CBC radio that if the tar ponds had been in British Columbia or Ontario, they would have been cleaned up long ago. Insurance executive Angus Ross could not contain his anger: "I'd like to see Jean Chrétien come down here and see his world's greatest country!"

Don Deleskie decided that he had had enough—he was fed up with the Joint Action Group. He was furious with the politicians and government bodies that had stood back mutely and watched the community be poisoned. He was expecially enraged that nothing was being done. The government was handing over $62 million in new funding, but none of the money was for relocating residents of threatened neighbourhoods, none was for compensating the families of dead and dying workers or members of the community poisoned by the steel mill and coke ovens—and even more incredibly, none of the new $62 million was for actually cleaning up the tar ponds.

Don announced to local media that on Monday, August 23, he would start the cleanup himself. He promised that he would take a shovel and a bucket and load the contaminated sludge into 100-gallon drums to get it away from the children in the community. Reporters and supporters gathered to watch as Don put his cane down and waded into the tar ponds from the Intercolonial Street side. No fences prevented access to the estuary. Any child could, and still did, lose toys by playing along its banks.

Don waded in the muck, stuck in his shovel and started digging. Observers, including one reporter, were overcome by the stench and fainted. Don kept scooping up the toxic water and sludge. "When a tree falls in the forest, you do not hear that, but when a member of this community falls, we hear it loud and clear," he said. "We're sick and tired of watching our brothers and sisters die."[10]

The provincial environment official in attendance told the press that the cleanup was underway. "It is a phased approach," he explained.

Don Deleskie is not a well man. His lungs are ravaged by years of living in the Pier and one year's work at the coke ovens. When asked whether he was not concerned for his own safety, he explained that he was prepared to sacrifice himself to protect the children. Just as he had done in his hunger strike in 1993, Don was prepared to put his life on the line to get action.

With reporters fainting on the sidelines, Don Deleskie started cleaning up the tar ponds. He is the only person to have made any progress in over 15 years.

From the very core of the Earth
comes a rumbling sound of woe.
Listen to what it says.
I have done you no wrong,
Why then do you poison me with your toxic waste of PCBs
destroying the food I so willingly give to you.
Take heed!
Soon I will be gone,
and you will cry.

—Shirley Christmas

CHAPTER TEN
Leaving Home

Frederick Street is all but abandoned now. The pattern of houses is even more of a patchwork. The windows are boarded over. On October 29, 1999, the first of the houses was demolished—the debris prevented by a labour dispute from being removed from the site. Louise Desveaux is still trying to find a home she can afford with the $34,000 government offer. She returned to her house on Frederick Street reluctantly. Daily she walks her German shepherd, Bear, past the empty homes of her neighbours. Several families defiantly remain, furious that Juanita McKenzie and Debbie Ouellette's efforts have, in their view, ruined their property values. As reporters gathered to watch the demolition, one angry resident, a lonely holdout, posted a banner proclaiming the street safe. Debbie Ouellette came to watch the first house come down. "After 15 years, I lost my home because of the environment," she said.[1]

Debbie and her family finally found a house they could afford. It is in Ashby, farther from the coke ovens, but not by much. She is not sleeping easily. Someday, she hopes, the whole neighbourhood where she now lives will be moved too. Anne Ross still lives with arsenic pooling on the basement floor and now with a strong chemical smell that she suspects is PAHs. No one will do any indoor air quality testing.

Juanita and Rickie are out of the fray, living on Sydney's outskirts in Howie Centre. Their loving relationship continues to give them strength, but Juanita has resigned from the JAG. Health is still their first concern. Although his respiratory problems dramatically improved after leaving Frederick Street, Rickie had another major heart attack on Christmas Eve 1999.

It is estimated that the plant had soaked up $3 billion in provincial and federal government support since it was "rescued" in 1967, and is now down to a workforce of 700. Despite efforts to sell the plant to Chinese and Mexican interests, and, most recently, a U.S.-based consortium, each sale has fallen through. It is likely the plant will be closed and sold for scrap metal in the near future.

The Joint Action Group has become even more fractious. Meetings now deal with appeals from ethics committee decisions. The ethics committee is developing its own jurisprudence, distinguishing cases of verbal abuse and threats from each other. In an appalling new chapter in JAG's history of expulsions, Mark Ferris has been sanctioned with a ten-month banishment. The "hearing" of his appeal took place in his absence. Mark Ferris faced an ethics complaint after his outburst at the summer's annual general meeting, when he accused the new chair, Dan Fraser, of being a government choice. The acting project manager, Germaine LeMoine complained about Mark's behaviour. Enraged that she had placed a flattering profile of Dan Fraser in the JAG newsletter just before the election, Mark argued that she had given Dan an unfair advantage over the other leading candidate, local lawyer and environmentalist Doug MacKinnon. The exchange became heated, and Mark called her a "red-haired bitch." Despite Mark's public apology, Germaine and another member of the support staff brought a complaint forward about his behaviour.

In what was rapidly descending into farce, Mark relied on the precedent of his own previous case, *Ferris v. Buchanan*, in which the ethics panel had ruled that Bucky Buchanan needed only to apologize for calling Mark an epithet and threatening him with a beating. The ethics panel chose to treat Mark differently and ruled that he should be suspended from the JAG process for ten months as punishment. Mark appealed and requested that the appeal take place at the next scheduled JAG meeting in November after his return from a business trip.

At the JAG round table meeting on October 27, the first items of business were two ethics appeals, one involving charges of industrial espionage and rumour-mongering between two other JAG members, and the other Mark Ferris's appeal against a ten-month suspension. The debate was bitter. Volunteer members of the JAG were deeply divided over the issue of delaying Mark Ferris's appeal. Even those who felt Mark deserved some punishment did not want to proceed in his absence. Dan Fraser had removed himself from the chair so that he could argue to proceed to expel Mark. The vote was very close. If government members present had abstained instead of voting to proceed with the appeal in the absence of the appellant, the hearing would have been postponed. Instead, the round table voted to hear Mark's appeal without waiting for Mark.

Nearly half of those at the meeting walked out as the appeal was conducted. To no one's surprise, it failed and Mark was expelled from the process. He has announced plans to launch a defamation suit against the JAG and the members who voted against him. It is astonishing that in 1999, a government-sponsored process could deny the most basic of rights—what lawyers call the right of fairness—by proceeding in the absence of the appellant.

In the wake of the meeting, JAG members were more polarized than at any other time in the process. John Steele, who left the table rather than participate in the appeal vote, told the media,

"It's the worst meeting I've seen in my life. . . . I don't think they [the JAG] can run a bakery."[2]

Ada Hearn watched from the observers' area in disgust. Describing the meeting as "pathetic," she bemoaned the lack of focus. "They're getting off base of what they're there for. I'm sick of hearing about personality conflicts. I have no hope at all that the sites will be cleaned up. It's all pretty sad."

JAG meetings now have police at the ready to keep out Mark Ferris—even from participating as an observer. The appearance of Bruno Marcocchio (who had not yet made his move to British Columbia) at a November 1999 meeting was enough to panic the chair into a rushed adjournment.

The promised "separation zone" around the coke ovens, which would remove more people from unsafe homes, has slipped to a hypothetical. A $57,000 contract was let to consultants to determine whether removing buildings on the site posed a health hazard justifying a separation zone. At an unruly public meeting in September 1999, the consultants explained that there might not be any separation zone at all. Meanwhile, the JAG's deadline for coming to a recommendation on what technology it favours for cleanup is 2002.

Of course, the cleanup plan will include pouring raw sewage into the harbour—bypassing the toxic estuary, which will still flow unimpeded into the sea. The children from neighbouring communities will be bused to the school that sits immediately above the old steel mill and its moth-balled but carefully maintained incinerator—a large segment of the community still hoping the incinerator will be fired up to get rid of the mess. The province's biomedical waste will be partially incinerated in the old municipal incinerator.

The Road Not Taken

There are technologies capable of cleaning up the tar ponds. Proposals sit unreviewed in the file drawers at the JAG offices. The most promising ideas involve ways of breaking down the very chemical bonds that hold PAH and PCB molecules together, rendering the resulting product as innocent as water and air. Processes now exist that release no emissions. Closed-loop systems, they are called.

The top three proposals could be selected within months and sent to a thorough environmental assessment under the federal environmental assessment law. With a public panel process, the whole community—not just its surrogate as represented by the JAG—could be involved. Remediating the coke ovens site is harder to imagine, as the toxic substances descend to depths of 24 metres. Some form of bio-remediation might work—allowing special strains of bacteria to "eat" their way through the toxics.

In the meantime, the mouth of the estuary should be temporarily blocked to prevent the daily flushing of toxic substances to the harbour. A sewage treatment system—preferably many small "living machine" systems—where the natural action of sunlight, living plants, and biological processes purify the water should be installed so that the dumping of raw sewage stops.

But, most urgently, the ongoing exposure to dangerous levels of chemical contamination must end. The most cost-effective way to do this is to relocate the community surrounding the plant. Given the rich cultural heritage of the area, ideally, neighbourhoods should be relocated intact. Any relocation must involve full community consultation and co-operation. But many streets and neighbourhoods may need to be moved. No one should spend every day of his or her life breathing toxic air.

Ideally, the Sydney tar ponds must provoke the same policy change in Canada that Love Canal did in the United States. After Love Canal, the U.S. government established the Superfund program, and while not perfect, it does relocate communities and clean up priority sites, however slowly. Canadians should not have to depend on the tender mercies of politicians and bureaucrats, laden with conflict of interest, deciding on an ad hoc basis whether communities should be moved. Such decisions require criteria and guidance within a national program. All the other toxic sites in Canada could then be dealt with as well. The most dangerous should be cleaned up first as we work our way through the legacy of governmental neglect and industrial irresponsibility.

Government should fundamentally restructure our current laws. Trading peoples' lives for corporate profits is a bad bargain. Many in the corporate world are already leading the charge, demonstrating the increased profitability of doing things right. Government coddling of poor performers only makes it more difficult for those ready to use closed-loop systems and reclaim all the by-products of their manufacturing process.

After 15 years of promises, the tar ponds are unchanged. People are dying. When will the nightmare end? The people of Frederick Street—and the people of Sydney—are waiting for an answer.

POSTSCRIPT

A Canadian Dilemma

It had taken years. Studies had piled upon studies. The committee of local citizens and government had worked with their experts to develop a consensus. However, the newly appointed provincial minister of the environment would not be rushed. Millions of dollars were at stake and he did not want to be pushed into a hasty decision. With Christmas holidays approaching, he gathered up the growing stacks of materials and took them home. He read every report, carefully weighing in his own mind the health risks faced by residents. Not everyone wanted to be moved. One could make a case in either direction.

Finally, he made a decision and signed the order to relocate the community and remediate the site. With luck, people could return to the decontaminated area someday. But for now, residents would receive compensation and be moved away from the toxic waste.

The year was 1985. The minister was not in the Nova Scotia government. The residents moved were not from Sydney. The community was Lasalle, Quebec, and the minister was Quebec's former minister of the environment, Clifford Lincoln.[1]

Lincoln struggled with the decision to relocate a community far less contaminated than Sydney. The area had been a dumping ground for wastes from oil companies outside Montreal, now in the riding of Finance Minister Paul Martin. With time, the remediation was completed, and houses once again sprang up in an area that had once been too toxic to allow residential use.

How can it be that in the same country of Canada, residents of one community can be so differently treated from residents of another? For years, Cape Bretoners have said that if the tar ponds were anywhere else, they would have been cleaned up long ago. The claim is more than hyperbole. If such a toxic nightmare had been in a First Nations reserve, or any other community lacking political clout it probably would not have ended either, unless there was the political will to do so. The common denominator in environmental protection in Canada is that there is no common denominator. When it comes to toxic waste sites, decision-making is ad hoc. If a community is fortunate enough to have a provincial minister of the calibre of a Clifford Lincoln, it may be relocated. But it is more a matter of luck than law.

Environmental Racism

When we look more closely at pollution of neighbourhoods in Canada and the United States a common theme emerges. Deadly toxic waste sites are more likely to be found near First Nations communities, or near non–First Nations communities of poorer people, people of colour and politically marginalized people. This phenomenon has been given a name: "environmental racism."

In the United States, the Environmental Protection Agency has created a large environmental justice program to address the problem. The statistics are overwhelming: more poor people, aboriginals and people of colour have their health compromised by pollution than wealthier, white communities. Sydney residents are right in believing that the conditions in which they live would not be tolerated in Toronto or Vancouver. There is no mystery about this reality—communities like Rosedale in Toronto, Rockcliffe Park in Ottawa, Westmount in Montreal or Shaughnessy in Vancouver are simply not threatened with dump sites.

First Nations communities are more exposed to environmental risk than other Canadians. The Lubicon First Nation in Alberta experienced dramatic increases in illness since their reserves were surrounded with sour gas wells. Within a two-year period in the early 1980s, over 400 oil wells were drilled within a 24-kilometre radius of the Lubicon village of Little Buffalo. UNOCAL, a California-based oil and gas company, built a sour gas battery plant within 5 kilometres of the same community. Sour gas is exactly what it smells like, and the rotten egg-smell of hydrogen sulphide emissions is posing health problems for the Lubicon.

Their health is also compromised by logging. As traditional sources of food in the bush are no longer available, the Lubicon are forced to rely on store-bought food. The rates of diabetes have soared, as have suicides and other social problems. What was once a healthy, self-sustaining community has been devastated. Currently, 95 per cent of the population is on welfare, while 35 per cent have health problems, ranging from tuberculosis to respiratory problems to cancer—at rates that exceed the national average.[2] Oil and gas development was rushed through approval processes before air emissions regulations could catch up.

In the 1940s, when the Inuit men of Deline in the Northwest Territories were recruited to mine uranium, they were never told that it was destined for nuclear weapons. When the bomb was dropped on Hiroshima, they never dreamt that they had unknowingly played a part. They were also never warned of any danger in carrying uranium out of the mine in sacks on their backs. As early as 1931, the government knew the dangers of the ore that the Inuit called "money rock." The men of Deline were carelessly, even criminally, exposed to excessive radiation from the uranium ore. The environment around them was also heavily contaminated.

Nearly 2 million tons of radioactive tailings were dumped into Great Bear Lake.

As the men died, survivors began to ask questions. The town of Deline, now known as the "Village of the Widows," is finally closer to answers—if no closer to justice. A federal Crown corporation knowingly allowed unprotected workers to be massively exposed to radioactive materials. The lake is still full of the 1.7 million tons of uranium waste dumped there. Survivors are demanding a health study, cleanup and compensation. An apology would also be in order, but so far it is the Inuit who have apologized. A delegation from Deline travelled to Hiroshima for commemorative ceremonies at the fiftieth anniversary of the bombing, August 6, 1999. They offered their heartfelt apologies for the part they unwittingly played in the annihilation of the city. Japanese citizens learned for the first time that the men who mined the uranium that killed hundreds of thousands had themselves paid with their lives.

Throughout the North, toxic chemicals have also affected the Inuit. Levels of exposure and accumulation of persistent toxic chemicals are at their worst in Inuit communities. The patterns of atmospheric cycling have made the North a dumping ground for industrial chemicals that were never used there. The Inuit diet is primarily from the wildlife of the North. The chemicals bioaccumulate, delivering a higher level of toxic concentration to each level up the food chain. As a result, the breastmilk of Inuit mothers is ten times as contaminated as that of southern Canadian women. In both the North and the South, mother's milk is so laden with toxic substances such as PCBs, DDT (and its breakdown product DDE) and lindane, that if it were offered for sale, it would be too contaminated to be approved as human food. Yet it is the only food that delivers to the newborn those immunities and antibodies that the growing child will need for life, and that may, in fact, be essential in developing an immune system to resist the effects of the same chemicals that contaminated the milk.

Toxic chemicals are indiscriminate and once in the environment they will eventually reach everyone. Each one of us carries at least

500 chemicals in our bodies that were unknown before 1920. After the discovery of chlorine during the First World War, the development of chlorine-based synthetic compounds launched a wave of new products, insecticides, plastics and chemicals of all kinds to which the human body has had little time to adapt. The American Environmental Protection Agency has stated the problem starkly: "Everyone in the U.S. has a body burden of dioxin reaching the potential for a national crisis."

In addition to their involuntary exposure through toxic waste sites, Canadians unwittingly dump poisons on their lawns. Of the 36 most common lawn pesticides, 34 are carcinogenic, 14 are suspected of causing birth defects, 21 damage the nervous system and 15 injure the liver and kidneys.

In total, between 1994 and 1996, 1.45 billion pounds of toxic substances were released in Canada, of which 280 million pounds were carcinogenic. We are poisoning our air, water and soil, somehow believing we will be immune from the inevitable effects.

Mapping the Hot Spots

"The legacy of 100 years of steelmaking" is the common phrase to describe the toxic mess known as the tar ponds and coke ovens in Sydney. With that description comes a subtle and misleading message that such disasters are historical—a thing of the past.

Sydney residents are still living with the ongoing health threat of the contamination, and the surrounding environment is still more polluted by government failure to clean up the tar ponds. Worse, the nature of government decision-making suggests that such disasters are not merely products of a time when we did not realize the consequences of industrialization. Instead, they reveal today's negligence and indifference. Despite a generation raised since the birth of the environmental movement and the creation of governmental departments charged with protecting the biosphere, Canada still lacks basic tools to protect human health and the environment and is allowing trade deals to undermine the tools we do have.

At the provincial and federal level, every environment department, whether federal or provincial, has seen its budget slashed by at least one-third in the last decade. Meanwhile, more responsibilities, including those for toxic waste sites, are being downloaded from federal to provincial responsibility.

Many communities across Canada have toxic waste sites, yet Environment Canada does not even attempt to maintain a list of such sites. From 1989 to 1995, a preliminary list was assembled but, before it could be organized into a proper inventory, budget cuts shut down the program. Why bother collecting the information when, unlike the United States, Canada has no national program for cleanup? The federal government has not even cleaned up those sites that were created by government itself, much less by its myriad of Crown corporations, such as those in the nuclear industry.

At best, there are estimates. The federal public accounts for 1999, tabled in the House of Commons in the fall of 1999, estimated the extent of liability of both government and corporations for contaminated sites at $30 billion.[3] There are an estimated 10,000 toxic sites across Canada—half of them on federal land.[4] The worst federal polluter has been the Department of National Defence, as well as former U.S. bases under DND responsibility. The Distant Early Warning System (DEW) line in Canada's North is dotted with toxic sites. DND sites are contaminated with fuel tanks, lead bullets, heavy metals and PCBs. In October 1999, Environment Minister David Anderson announced intentions to develop a plan to deal with contaminated sites, but only on federal lands.[5]

When there is a plan, much progress occurs. While it is a long way from being fully implemented, the 1987 Great Lakes Water Quality Agreement did result in significant cleanup efforts. Unlike the tepid language of the governmental memorandum of understanding for the Muggah Creek estuary, the goals of the agreement were ambitious and stated in clear, unequivocal language. The agreement affirmed the commitment of the governments of Canada and the United States, as well as the governments of

Ontario and certain U.S. states, to swimmable beaches, edible fish and drinkable water. Sadly, the sludge on the floor of our largest freshwater lakes will remain contaminated with PCBs and other persistent bio-accumulative toxic substances for a long time. Scientists fear that disturbing them in a cleanup could worsen the problem.

Another steel town, Hamilton, Ontario, has seen millions of dollars in Canada and U.S. funding spent on cleanup. Hamilton Harbour was extensively contaminated but it was made a priority under the Great Lakes Water Quality Agreement. Collins Bay within the harbour has been sufficiently remediated to be removed from the list of areas of concern. But, even though substantial progress has been made, the area still has many toxic hot spots. Randle Reef in Hamilton Harbour was badly polluted by Stelco, and the sediment is highly contaminated, primarily with PAHs. The cleanup on Randle Reef has yet to begin.

From the Great Lakes flows the St. Lawrence River, heavily contaminated by chemical industries on both sides of the border. In upstate New York, industry practices led to the much-publicized Love Canal disaster where Hooker Chemical had abandoned a toxic site upon which a large housing subdivision was built. Lois Gibbs and a dogged band of residents fought for and eventually won a massive relocation of the whole community, leading, as we have noted, to the U.S. government's creation of the Superfund program to remediate toxic areas.

The Canadian side of the border also has its horror stories. The once pristine and productive lands of the area now known as Cornwall, Ontario, had supported the Mohawk people well for generations. They were successful farmers, with fruit trees and market gardens, enjoying a prosperous life on the banks of the St. Lawrence. Henry Lickers, the head of the Akwesasne environmental department, speaks eloquently of all that was lost by a people whose land was taken by settlers in Upper Canada and then eventually poisoned by industrialization.

The radioactive hazard that devastated Deline in the Northwest Territories is not restricted to Canada's North. As Anne McIlroy noted recently in the *Globe and Mail,* "If Deline was the first link in Canada's nuclear chain, Port Hope, a pretty town on the shore of Lake Ontario, was the second."[6] The uranium ore from the Eldorado mine in Deline was loaded onto boats, shipped down the Mackenzie River and ultimately transported to Port Hope, where Eldorado also ran the refinery for radium and uranium. The refinery, which had begun operating in the 1930s, became a Crown corporation in the Second World War for security reasons. Port Hope was heavily polluted with radioactive and other toxic wastes, such as arsenic, fluorine and lead. Contaminated fill was used for the foundations of buildings, and basements became dangerously radioactive with radon gas.

In the mid-1970s, St. Mary's primary school was closed due to high levels of radon gas. Millions were spent removing radioactive fill and hundreds of properties were de-contaminated. But radioactive contamination continued to take its toll. Residents such as Pat McMillan complain of higher cancer rates and have demanded a long-term health study. "My brother, sister and I thought that everywhere in the world there must be thick yellow smoke bellowing out of smokestacks, smelling up the air with the smell of rotten eggs," Pat McMillan told the Atomic Energy Control Board, "It was normal for us to rub our eyes because of the smoke that would drop its dust particles to the ground, coating everything it touched with it."[7]

Working with residents, epidemiologist Dr. Trevor Hancock of York University has proposed a study to track increased levels of cancer, asthma, emphysema, thyroid problems, learning disabilities and other ailments that may be linked to the radioactive contamination. "After all these years, no one has ever drawn a blood sample. I feel the people of Canada owe this community a study," Dr. Hancock told the *Globe and Mail.*[8]

The daughter of a prominent activist for the cleanup of Port Hope, Molly Malloy, is also demanding a health study. At 40 years old, she has been diagnosed with a brain tumour and told she has less than two years to live. "We are in the 1990s. We are a wealthy country, and industry has a responsibility to protect the health of the people who live nearby," she said. "If they won't do the most basic health study, what does that say?"

Other areas in Ontario adjacent to uranium mining and nuclear reactors are also contaminated. For example, nearly 200 million tons of uranium tailings are piled up across areas such as Elliot Lake in northern Ontario. Downstream from Chalk River where Atomic Energy of Canada Ltd. operates a research reactor, the one a young Jimmy Carter visited during a nuclear accident, the plume of radioactive contamination moves steadily down the Ottawa River to the nation's capital.

Toxic Dumps Coast to Coast

Apart from the obvious problems in Deline, Port Hope and areas around the Great Lakes and St. Lawrence, many other communities are exposed to toxic threats. Abandoned Domtar plants continue to pollute. Domtar's wood preservative plant in Newcastle, New Brunswick, operational since 1924, was shut down after it was found to have contaminated the municipal water supply. The site was put on the province's list of toxic sites needing cleanup back in 1988, but nothing has been done to date.

A similar problem plagues the old Domtar wood preservative site in Transcona, Manitoba, to the east of Winnipeg. In the spring of 1999, 35 homes backing onto the old site were to have their backyards excavated after creosote-contaminated soil was discovered 1.5 metres below the ground. Domtar had manufactured railway ties and fence posts on the site from 1911 to 1976 when the environment minister of the time, Gary Filmon, gave the company the green light to sell the land to a developer. The developer went bankrupt and the land reverted to Domtar. Residents complained about the foul smell for years before Manitoba declared the site an environmental hazard in 1981. It took until 1992 for Domtar to do anything, and that was only after the province threatened to do the cleanup itself and charge Domtar. Now the residential cleanup has begun with a commitment to return the properties to their pre-excavation state. The Domtar site is supposed to become a bio-preserve run by the Fort Whyte Centre, with prairie grasses and wildflower gardens.

In Alberta, Domtar's old creosote site in the town of Cochrane west of Calgary is one of many the company has in western Canada. Domtar is trying to remediate it and develop it into a commercial and minor residential area. The Cochrane Environmental Action Committee has established a citizens' watchdog committee to keep the developer honest.

Squamish, B.C., is home to the single largest source of heavy metal pollution on the North American continent. The old chloralkali plant on the Squamish River estuary is a mega-source of mercury pollution. The site itself is toxic and some mercury still leaches into Howe Sound and surrounding areas. Sediments off site in the estuary area are contaminated although studies to date indicate the mercury has been covered with sediment. Due to airborne discharges when the plant was in operation, some of the surrounding land areas also indicate levels above the acceptable standards. A sampling program is presently underway to determine the extent of contamination. Literally tons of mercury a year were discharged during operation of the plant.

Mercury has a long history as a dangerous pollutant. The late Warner Troyer, arguably Canada's first environmental journalist, documented the appalling, even criminal devastation of the White Dog Reserve downstream from the Reed Paper Company in Dryden, Ontario.[9] Mercury contamination from enormous hydroelectric facilities also deprived the James Bay Cree of their traditional diet. While some mercury pollution has been dealt with, the increased burning of coal in the Ohio Valley and in Ontario is causing increased levels of mercury in the rain and is contaminating wildlife to dangerous levels, particularly in eastern Canada.

Abandoned mine sites are also common sources of pollution. One of the worst in Canada is the former Britannia mine site in British Columbia's Howe Sound, already contaminated by mercury poisoning from the old chloralkali plant. The old mine, 10 kilometres south of the town of Squamish, emits a constant flow of extremely high levels of copper into Howe Sound. The toxic copper is damaging fish habitat and the entire mountain is filled with toxic acid mine drainage. In fact, Britannia Creek is so toxic that the creek water itself kills fish. The poisonous discharges from this site have been flowing for decades.

To the north is the massive pollution legacy of the bankrupt Royal Oak mine in Yellowknife, Northwest Territories, where fifty years of gold processing left more than 250,000 tonnes of arsenic trioxide inside the mine's caverns. To prevent the toxic waste from reaching aquifers, a pumping system must labour 24 hours a day. Cleanup costs are estimated at $250 million.

In Belleville, Ontario, another mine site is contaminating the surrounding area. The Deloro mine north of Belleville is discharging large amounts of arsenic and other heavy metals into an important fish habitat in the Moira River. Charges against the province have been laid under the provincial Environmental Protection Act for allowing dangerous levels of radiation to be discharged near the Deloro site.

PCB storage is also a toxic threat. On military bases, old warehouses, and discarded transformers contain PCBs. As we saw in Sydney, PCBs were once regarded as benign, and dumped carelessly. Now, they are ubiquitous in our environment and our bodies. Although PCBs are one of a handful of regulated chemicals under the Canadian Environmental Protection Act, regulations to ensure safe storage and handling of PCBs are inadequately enforced.

Meanwhile, old city dumps can carelessly dispose of toxic substances without any thought to the environment. In Kingston, Ontario, the Belle Park dump, formerly a wetland on which the city dump was located, sits at the edge of the Cataraqui River. Although the dump was closed in 1971, PCBs and over 100 other chemicals in the leachate continue to enter the Cataraqui River. The entire site (soil, vegetation, groundwater aquifers) is contaminated with PCBs. Sierra Legal Defence Fund laid criminal charges against the city under the federal Fisheries Act and charges by the provincial Ministry of the Environment followed. In December 1998, the charges were successfully prosecuted with convictions against the city. The case is currently under appeal.

Another dump near Kingston is also a major polluter. The Storrington dump, located approximately 15 kilometres north of Kingston on the Rideau Canal, is owned by Canadian Waste Systems (formerly Laidlaw and originally Tricil).Though closed since 1991, the leachate plume continues its journey toward the Rideau Canal on two sides. Residents' drinking wells are now beginning to show signs of possible contamination. In total, there are over 2,000 closed dumps in Ontario, most of which are probably leaking hazardous waste into the environment. Sierra Legal Defence Fund has taken action in three cases and is investigating similar sites in Picton, Belleville, Brockville, Hamilton and London.

Dumps in New Brunswick have also been identified as toxic threats. The Howe's Lake and French Village dumps, in the Saint John area, have been identified as the most threatening of the closed dumps in the province. Toxics buried there are leaching into the Saint John River system.

Memories are dim about the toxic contamination around the military base at Gagetown, New Brunswick. Before the creation of the notorious defoliant "Agent Orange" used in the jungles of Vietnam, various proposed formulations were tested at Gagetown, including defoliants Agent White and Agent Purple. Virtually the entire province was exposed to contamination. From 1952 to 1994, New Brunswick had its own air force spraying poison to combat the spruce budworm. No analysis has ever been done of the long-term residual toxic contamination from air bases, empty barrels of chemicals and spraying of forests.

Who Protects Public Health and the Environment?

The Canadian Environmental Protection Act (CEPA) is the most important piece of legislation in Canada for the regulation of toxic chemicals. It is administered jointly by the ministries of environment and health and operates on a chemical-by-chemical basis.

Between the spring and fall of 1999, the Prime Minister's Office forced the passage of amendments that weakened and redrafted CEPA to satisfy the industry lobby. Originally passed in

1988, the act cobbled together various bits, such as the Ocean Dumping Act, pre-existing regulations to control nutrients in water and, its primary basis, the Commercial Chemicals Act. From the very beginning it failed to live up to its name—excluding large classes of toxic substances of concern to Canadians. If a chemical is so toxic that its primary purpose is to kill things, then CEPA only operated if the chemicals were left lying around. As long as the chemicals were being widely dispersed over the environment, they fell under the Pest Control Products Act. If the toxic material also happened to be radioactive, CEPA would not apply.

Still, CEPA did promise comprehensive management of regulated substances—from "cradle to grave" as the press release at the time of passage touted. The problem is that studying and listing toxic chemicals, one substance at a time, is a long and difficult process. The deficiencies in Canada's toxic chemical management were set out in the 1999 report of the Commissioner for Environment and Sustainable Development, Brian Emmett. In his view, the system was so flawed that the health of Canadians was at risk. Of the 23,000 toxic substances in Canada, only 31 have been subjected to a conclusive review process. The Priority Substances List, which was intended to fast-track the operation of CEPA, has been an exercise in slow motion.

Commissioner Emmett's report also highlighted the lack of information-sharing between departments as was evident in the tar ponds disaster. The shared role of environment and health ministers should have worked to integrate our life support systems—the air we breathe, the water we drink, the food we eat—with our state of health. It did not. Instead, unsafe chemicals have remained in use while health and environment bureaucrats engage in turf warfare.

The only good news in the last ten years has come from the Supreme Court of Canada. It reconfirmed the importance of the federal role in protecting public health and the environment from poisonous substances.

After heroic efforts by the House of Commons environment

committee to improve the act through its mandatory five-year review, the industry lobby went to war against the committee. It was the worst, most protracted and unpleasant parliamentary process of any bill in memory.

When the all-party committee presented its report, the chemical and aluminum industries used every weapon in their arsenal to gut the bill. The CEO of Alcan Aluminum Ltd, Jacques Bougie, wrote to Prime Minister Jean Chrétien, warning him that if the bill was enacted into law as written, "it could force the closure of all aluminum smelters in Canada."[10] It was not necessary for Mr. Bougie to remind Mr. Chrétien that one of the threatened smelters was in the prime minister's Shawinigan riding. The source of Bougie's concern was that the bill might be used to regulate PAHs. Already listed on the toxic substances list, PAHs have been under a special review within CEPA for the last five years to determine the appropriate approach to regulation. The multi-stakeholder advisory committee includes industry, of course, and consensus has not been possible.

Some of the very same poisons that poured out of the coke ovens are also emitted from aluminum smelters. Benzopyrene turns up in the St. Lawrence downstream from smelters, and according to Alcan's vice-president, "There is no smelting technology that does not emit a detectable, almost negligible, level of PAHs."[11] Thus, when beluga whales wash up dead in the St. Lawrence River, their flesh must be treated as hazardous waste.

The industry focused on CEPA's new goal of the "virtual elimination" of certain inherently toxic and bioaccumulative chemicals, deciding that the concept had to be rejected. Such lobbying was not new. In response to industry pressure a year before, Environment Canada staff had prepared an analysis of wording changes demanded by the aluminum industry. The memo noted that the language demanded by industry would "create an internal contradiction that would make virtual elimination impossible." Yet the bill that was brought before the House for passage had nearly identical language to that rejected a year before as unworkable.

The result was predictable. Key sections of CEPA were made

unintelligible. The drafting was incomprehensible. It moved tentatively towards the possibility of virtual elimination, but failed to adopt a goal to achieve it. The industry also demanded that the precautionary principle, which requires that actions to protect public health and the environment not await 100 per cent proof when caution would require action, also be gutted. The bill sent to the House was amended to require that actions of a precautionary nature only be allowed when "cost-effective." Nothing in the new "improved" CEPA creates an imperative to clean up toxic waste sites—or even to catalogue them.

CEPA was so badly damaged by the prime ministerial cave-in to industry that a most extraordinary parliamentary rebellion took place. The three members of parliament within the governing Liberal Party, those most knowledgeable about the bill, voted against it. Charles Caccia, a former minister of the environment under Pierre Trudeau, was chair of the House of Commons committee that had dedicated years to the review and amendment process for CEPA. Joining him in rejecting the bill were Karen Kraft-Sloan, an Ontario MP and former parliamentary secretary to the environment minister who had also worked on the House Committee process, and Quebec's former environment minister—the same provincial minister who had relocated the contaminated community within Lasalle, Quebec—who was now a federal politician, Clifford Lincoln.

All three MPs believed the government had so emasculated the bill that it was now worse than the version of the bill passed in 1988. Efforts focused on the Senate to improve the bill, with progressive positions adopted by Tory senator Mira Spivak and committee chair Ron Ghitter. But the Prime Minister's Office and the Senate Liberal leader pushed all Liberal senators hard to approve the bill without changes. The lack of enthusiasm for the task was evidenced in a report that accompanied the bill to the Senate floor. The Liberal majority of the Senate committee urged that upon passage, the bill should be subjected to an immediate review to deal with its failings.

Toxic Dumping

While corporate lobbyists successfully gut our legislative tools, the potential profits from toxic waste are eroding our standards. The economic benefits of becoming a toxic dumping ground are beginning to change Canada's reputation. We will accept PCBs for disposal and incineration in Canada. The United States will not. We are pushing hard for the right to be the long-term repository for highly radioactive plutonium, removed from the warheads of U.S. and U.S.S.R. missiles. The campaign originates not with the former Soviet Union or the United States, but within Canada from the massively subsidized Crown corporation, Atomic Energy of Canada, Limited (AECL), supported by this country's biggest booster of nuclear energy, Prime Minister Jean Chrétien. Expert analysis of the disposal methods for plutonium favours leaving the plutonium where it is, and then encasing it in glass—a process called vitrification. Vitrification wins on every point as the least expensive, and safest, from an environmental and security viewpoint. But Chrétien and AECL are desperate to establish a global trade in plutonium waste so that AECL will have a long-term contract for disposal.

The open door policy to hazardous waste is not restricted to federal agencies. Ontario's Harris government has drastically increased the importation of toxic waste to the province. From 1997 to 1999, the province had asked the federal environment department to accept all hazardous waste applications from U.S. companies wishing to dump materials in Ontario. The blanket approvals letter was recently revoked by the province's new environment minister, Tony Clement. Meanwhile, in the first six months of 1998, nearly 11 million tonnes of hazardous materials were shipped into Ontario from the United States.[12]

Industry analysts know why companies are prepared to ship materials hundreds of kilometres from the source to ultimate disposal in Canada—our regulations are lax, our costs are lower, and there is less chance of being sued or prosecuted.

Trading Away Environmental Protection

The federal government continues to sign international trade agreements that prohibit all levels of Canadian government from passing legislation to protect the environment. The sad story of MMT shows just how much control the federal government has given away through such agreements.

In 1997, the federal government took the unusual step of banning a persistent neurotoxic substance used as an anti-knock agent in gasoline. The move was unusual because the government rarely bans a toxic chemical in Canada, tending instead to treat toxic chemicals as though they had constitutional rights—innocent until proven guilty. In fact, the banning of MMT was an appropriate use of the "precautionary principle." The science on the key ingredient in MMT, manganese, is well established. Manganese in occupational exposure can lead to a disease called "manganism" which closely resembles the tremors and nervous system breakdown of Parkinson's disease. Manganese exposure can cause a progressive deterioration of the brain. This condition is particularly dangerous for older people, leading to premature and accelerated aging of the brain.

MMT was introduced in the early 1970s as a gasoline additive by the same company that had manufactured and sold leaded gas, Ethyl Corporation of Richmond, Virginia. After decades of defending leaded gas as a safe product, Ethyl Corp. knew the jig was up. Leaded gas was on the way out, and the company wanted to protect its prime business as a manufacturer of gasoline additives. Ethyl Corp. began flogging MMT as a replacement for lead in gas.

The U.S. Government soon rejected registration of MMT. Canada, relying on the same data, decided it could be registered for use. Health Canada did note, however, that there were significant data gaps about how MMT might affect vulnerable groups, such as children, pregnant women and the elderly.

By the 1990s, the automakers were complaining to the government about MMT. They said that it was gumming up the on-board diagnostic systems of cars, compromising the air pollution control devices. The Big Three car manufacturers became very concerned about the financial repercussions of violated warranties if MMT reduced the effectiveness of catalytic converters. Pressure to ban it was exerted on Environment Minister Sheila Copps. Environmental and health groups supported the ban, arguing that MMT not only increased air pollution, but could poison the brains of Canadians. Twenty years after registration, Health Canada had done nothing to deal with the "data gaps" around its health impacts.

Laboratory studies suggested that MMT could provoke increased aggression in animals, as well as create symptoms that could be described as attention deficit disorder if they occurred in humans. The experience with leaded gas had proven that if you wanted to introduce a toxic heavy metal into the blood and brains of children, then adding it to gasoline was a good delivery mechanism.

In the spring of 1997, MMT was finally banned in Canada. A decade earlier that would have been the end of the matter. But Ethyl Corp. did not accept the regulatory decision. Now it had recourse through the North American Free Trade Agreement (NAFTA) to challenge Canada's decision. In fact, under provisions of Chapter 11 of NAFTA, companies from one of the three NAFTA countries who lose profits based on a regulatory decision in one of the other countries can sue the government for damages. Thus, Ethyl Corp, as a U.S.-based company, was able to sue the government of Canada for banning its neurotoxic gasoline additive, claiming damages in the amount of $350 million (Canadian) for lost profits and damage to its reputation. The hearings would be completely private. A three-person arbitration panel would hear the arguments of the Canadian government and Ethyl Corp. No independent scientific briefings would be allowed by environmental or health groups.

Under NAFTA Chapter 11 it didn't really matter if the government had been right to ban MMT or not. The real issue was whether Ethyl Corp. had rights to profits "expropriated." As Ethyl Corp.'s Canadian lawyer Barry Appleton has said, it wouldn't matter if a substance was liquid plutonium destined for a child's breakfast cereal. If the government bans a product and a U.S.-based company loses profits, the company can claim damages under NAFTA.

In the summer of 1998, the government caved in. It withdrew the regulation removing MMT from use, paid Ethyl Corp. $19 million as compensation for its "trouble," and issued a public apology in which Christine Stewart, then minister of the environment, explained that the government had never had adequate grounds to ban MMT. A spokesperson for Ethyl said at the time, "It's a very happy day, a significant step for Ethyl Corp. and its business worldwide."

Within days of the MMT settlement, Barry Appleton filed a claim for another U.S.-based corporation, S.D. Myers of Ohio, a company in the business of hazardous waste. It wanted PCBs from Canada for disposal in its Ohio plant. Myers claimed damages from a nine-month-long ban against the export of PCBs to the U.S. from Canada, initiated by Sheila Copps, and upholding the principles of the Basel Convention on hazardous wastes. The ban was removed quickly under a NAFTA threat. Even though Sierra Club in the U.S. had successfully sued to prevent Canadian PCBs from entering the U.S., it was still possible for S.D. Myers to sue for damages on a moot point.

Other cases are piling up: from a company wanting to export water from B.C., from a wood product company claiming that Canada's forest export rules deprive it of access to forest products, and even from a Canadian company, Methanex, challenging a California ban on a carcinogenic gas additive, MTBE.

The Chemical Industry's Lobby

The world's chemical lobby is big business. International Council of Chemical Associations (ICCA) members account for about 80 per cent of the $1.6 trillion in annual world chemical industry production. In the *International Trade Reporter*, the ICCA boasted that world trade in chemicals is second only to automobiles, in terms of global trade in manufactured goods. Of course, the council isn't counting the trade in armaments, which still tops them all.

Even with the power the chemical lobby already wields, it wants to be more powerful still. Meeting in Geneva in June 1999, the ICCA agreed to push for zero tariffs on chemicals worldwide by 2010. The effect of tariffs is to slow down the trade in a product. Some tariffs have been rightly attacked as unreasonable domestic trade barriers. But industry wants to see all such barriers removed in a number of sectors—fishery products, forest products and now chemical products.[13] Moreover, the global industry wants other non-tariff barriers—described as import licensing, quotas, dual pricing, and discriminatory standards, removed as well. Environmental regulations can also be considered a non-tariff barrier.

"Our job now is to get governments to agree to this goal," ICCA Chairman Gerry Finn said. "The chemical industry is united around this objective. We want to get governments similarly united." The ICCA wants the World Trade Organization, already famous for successfully challenging a number of domestic environmental regulations around the world, to enforce the global free trade in chemicals. Canada appears to have no objections.

Lessons Unlearned

Here, then, is the greatest tragedy of the Sydney tar ponds story. It would appear that the residents of Sydney, and especially the residents of Whitney Pier and Frederick Street, have suffered and fought in isolation and perhaps in vain. We appear not to have learned one thing from their ordeal. We continue to talk about the "trade-off" between jobs and the environment; jobs and health. The tar ponds saga should have taught us that such trade-offs are wrong economically, environmentally and morally.

There are now several decades of documented proof of the deep harm done to humans, other species and the earth by the noxious and toxic chemicals so cavalierly dumped into open waterways around the Sydney steel plant and at other poisoned sites right across Canada. Yet governments and many corporations continue to turn a blind eye to the clear and present danger these chemicals pose to the very future of humankind. In the end, our collective failure to learn from their suffering may be a greater offense to the courageous people of Sydney than the toxic site itself.

It is never too late to change our behaviour, our values and our laws. But it is late in the day. The sun is setting on a chemical-safe world; Sydney, Nova Scotia, has sounded the alarm. Will we hear it in time?

Chronology

1900	Dominion Iron and Steel Company (DISCO) begins construction of steel plant in Sydney
1910	Amalgamation of DISCO and Dominion Coal Company into Dominion Steel Corporation
1920	British Empire Steel Company (BESCO) created through corporate merger of Dominion Steel Corporation, Nova Scotia Steel, the Halifax shipyards, Wabana iron mines and the Eastern Car Company
1928	Creation of DOSCO—Dominion Steel and Coal Company—controlling the steel mill
1957	DOSCO becomes a subsidiary of British multinational, Hawker-Siddeley
October 13, 1967	"Black Friday," the day Hawker-Siddeley announced its intentions to close the steel mill
November 20, 1967	Parade of Concern to save the steel mill

December 1967 Creation of Sydney Steel Corporation (SYSCO), a provincial Crown corporation, taking responsibility for the steel mill

1982 The federal Department of Fisheries and Oceans closes the harbour to lobster fishing due to contamination of lobsters

1986 First announcement of federal-provincial funding for clean-up of tar ponds with promised completion date of mid-1990s

September 4, 1994 N.S. government assumes ownership of incinerator through Crown corporation, Sydney Tar Ponds Clean-Up Inc. (STPCUI)

January 1996 Province admits incinerator will not work; announces Plan B, the "encapsulation option" leaving the toxic sludge where it is and burying it in slag

August 1996 Province abandons encapsulation option,; federal-provincial announcement of new process to determine cleanup approach—the Joint Action Group (JAG)

Spring 1998 Ooze appears in brook along Frederick Street

May 1999 Evacuation of residents of Frederick Street and Curry's Lane

June 1999 Federal-provincial announcement of $62 million of JAG process, health studies, demonstration projects and sewer collector pipe; residents of Frederick Street and Curry's Lane to be offered permanent relocation

Endnotes

Chapter One: Paradise Lost

[1] Robert Pichette, trans., "Gobineau's Portrait of Sydney, 1859," *Cape Breton's Magazine* 53. First published in Gobineau's *Voyage à Terre Neuve.*

[2] E.R. Harvey, *Sydney, Nova Scotia: An Urban Study* (Toronto: Clarke and Irwin, 1971).

[3] J. McMullen and S. Smith, "Toxic Steel: State-Corporate Crime and the Contamination of the Environment," in *Crimes, Laws and Communities,* ed. D. Perrier et al (Halifax, N.S.: Fernwood Publishing, 1997).

[4] P. MacEwan, *Miners and Steelworkers* (Toronto: A.M. Hackert Ltd., 1976), 7.

[5] J. Murphy and P. MacNeil, "Legacy," a paper of the Beaton Archives, 1991, University College of Cape Breton.

[6] M.R. Campbell, "The History of Basic Steel Manufacture at Sydney, Nova Scotia," *Canadian Mining and Metallurgical Bulletin* 45, no. 42 (1952): 331–39, 1952.

[7] For the following description of the steelmaking process, the authors are

indebted to: former steelworkers John Murphy and Paul MacNeil, who wrote a detailed and understandable introduction to the process, entitled "Steelmaking," as part of the Beaton Institute's "The Steel Project" (July 5, 1991, Beaton Archives, University College of Cape Breton); *Making Steel*, video co-produced by the National Film Board of Canada and the Beaton Institute 1996; and Dan Yakimchuk, who clarified many details.

[8] Campbell, "The History of Basic Steel Manufacture."

[9] Murphy and MacNeil, "Legacy."

[10] Campbell, "The History of Basic Steel Manufacture."

[11] Ibid.

[12] McMullen and Smith, "Toxic Steel."

[13] D. Chisholm, "History of the Sydney Steel Corporation," prepared for the *Muggah Creek Watershed Report* (University College of Cape Breton: The Beaton Institute, 1997).

[14] William N.T. Wylie, "Coal Culture: The History and Commemoration of Coal Mining in Nova Scotia," a study prepared for the Historic Sites and Monuments Board of Canada. Autumn 1997.

[15] McMullen and Smith, "Toxic Steel."

[16] J.R. Bishop "The Sydney Steel Plant—Government Policy and Public Ownership," master's thesis cited in Chisholm, "History of the Sydney Steel Corporation."

[17] M. Katz and R.D. McKay, "Report on Dustfall Studies at Sydney, Nova Scotia: Analysis and Distribution of Dustfall in the Sydney Area During the Period February 1958 through September 1959" (Ottawa: Health and Welfare Canada, 1959).

[18] Ibid.

[19] Ibid.

[20] Arthur D. Little, "The Future of Steel Making in Sydney: Report to the Government of Nova Scotia" (Halifax: Government of Nova Scotia, 1960), 3–4.

Chapter Two: Sons of Steel

[1] *Making Steel*, video documentary co-produced by the National Film Board of Canada and the Beaton Institute of the University College of Cape Breton, 1996.

[2] George MacEachern, *An Autobiography: The Story of a Cape Breton Island Radical* (Sydney, Nova Scotia: University College of Cape Breton Press, 1987).

[3] Ron Crawley, *Off to Sydney: Newfoundlanders Emigrate to Industrial Cape Breton, 1890–1914, Acadiensis: Journal of the History of the Atlantic Region* (spring 1988).

[4] Ibid.

[5] Elizabeth Beaton, *Housing, People and Place: A Case Study of Whitney Pier* (University College of Cape Breton: Beaton Institute, 1996).

[6] Ron Crawley, *Off to Sydney.*

[7] Syd Slaven, "Safety and Accidents at the Sydney Steel Corporation, 1901-1997," (master's thesis, University College of Cape Breton, 1997).

[8] Elizabeth Beaton, *From the Pier Dear! Images of a Multicultural Community* (Sydney, Nova Scotia: Whitney Pier Historical Society, 1993).

[9] Ron Crawley, "Class Conflict and the Establishment of the Sydney Steel Industry, 1899–1904," in *The Island: New Perspectives on Cape Breton's History, 1713–1990*, ed. Kenneth Dunevin (Fredericton, N.B.: Acadiensis Press, 1990).

[10] George MacEachern, *An Autobiography.*

[11] Frank Smith, *Brief History of Local 1064 United Steelworkers of America and Its Achievements*, AFL-CIO Publication, 1985.

[12] Ron Crawley, "What Kind of Unionism: Struggles Among Sydney Steel Workers in the SWOC Years, 1936-1972," *Labour/ Le Travail*, 1997.

[13] Nelson Muise, interview by authors, 16 February 1999.

[14] Pat Wall, Harry Muldoon, interview by authors, 26 February 1999.

[15] Barb Lewis, interview by authors, 18 February 1999.

[16] Ed Johnson, interview by authors, 18 January 1999.

[17] Clyde Hoban and Don Puddicomb, interview by authors, 26 February 1999.

[18] Syd Slaven, "Safety and Accidents."

[19] Donnie MacPherson, "Tar Pond Tango," *New Maritimes Magazine* (April 1999).

[20] Donnie MacPherson, interview by authors, 27 February 1999.

[21] Steve McLeod, "Cancer ravages industrial Cape Breton," Canadian Press, 6 January 1995.

[22] Joan Bishop, "Sydney Steel: Public Ownership and the Welfare State, 1967-1975," in *The Island: New Perspectives on Cape Breton's History, 1713–1990*, ed. Kenneth Dunevin (Frederiction, N.B.: Acadiensis Press, 1990).

[23] Michelle Gardiner and Valda Walsh, “No Smoke, No Baloney” (master’s thesis, University College of Cape Breton, 1992).

Chapter Three: In the Shadow of the Valley

[1] Elizabeth Beaton, *Housing: People and Place. A Case Study of Whitney Pier* (University College of Cape Breton: Beaton Institute, 1996).

[2] Ibid.

[3] Ibid.

[4] Dan Yakimchuk, interview by authors, 28 February 1999.

[5] Elizabeth Beaton, *Housing.*

[6] Ibid.

[7] Clotilda Yakimchuk, interview by authors, 27 February 1999.

[8] Peggy and Eric Brophy, interview by authors, 27 February 1999.

[9] Indian Reserve No. 28, Kings Road Statement of Fact.” A statement of fact on Membertou’s history researched and compiled by staff at the Mi’kmaq Treaty and Aboriginal Rights Research Office, Sydney, N.S., 1999.

[10] John Campbell, “Long road to establishing native community,” *Cape Breton Post,* 29 November 1997.

[11] Letter to the Nova Scotia Inspector of Indian Agencies from Duncan C. Scott, Deputy Superintendent General, Indian Affairs in Ottawa, dated 26 July 1920.

[12] Letter to Indian Affairs in Ottawa from A.J. Boyd, Nova Scotia Indian Superintendent, dated 15 July 1925.

[13] Geoffrey York, *The Dispossessed: Life and Death in Native Canada* (Toronto: Lester & Orpen Dennys, 1989) p. 64.

[14] Shirley Christmas, interview by authors, 26 February 1999.

[15] Geoffrey York, *The Dispossessed*, p. 66.

[16] Shirley Christmas, Remarks to a Public Lecture, reprinted in the *Chronicle Herald,* 8 June 1996.

Chapter Four: State-Sponsored Crime

[1] J. McMullen and S. Smith, “Toxic Steel: “State-Corporate Crime and the Contamination of the Environment,” in *Crimes, Laws and Communities*, ed. D. Perrier et al (Halifax: Fernwood Publishing, 1997).

[2] E.J. Kilotat and H.J. Wilson, "An Evaluation of Air Pollution Levels in Sydney, Nova Scotia," internal report, Health and Welfare Canada, Environmental Health Directorate, Air Pollution Control Division, 1970.

[3] Donnie MacPherson, interview by authors, 27 February 1999.

[4] Interview with Warren Gordon, now a noted photographer in Sydney, by author, 28 February 1999.

[5] Donnie MacPherson, interview by authors, 27 February 1999.

[6] Ibid.

[7] V. Havelock, "Air Pollution Assessment of Sydney Steel's Present and Future Steel Making Operations, Sydney Nova Scotia," internal report, Environment Canada, August 1973.

[8] Ibid., vii.

[9] Ibid., 6.

[10] Ibid., viii.

[11] M. Earle, "Sydney Steel Plant—A Chronology," Beaton Archives, University College of Cape Breton.

[12] P. Choquette, "Air Pollution Assessment of Sydney Steel's Present and Future Coke-Making Operations, Sydney, Nova Scotia," internal report, Air Pollution Control Directorate, 1974, marked "Restricted."

[13] Ibid. 25.

[14] M. Grimard, "Report of the Sydney Respiratory Health Survey," Environmental Health Directorate, Health Protection Branch, Department of National Health and Welfare, March 1977, 51.

[15] Ibid.

[16] McMullen and Smith, "Toxic Steel."

[17] L.P. Hildebrand, "Environmental Quality in Sydney and Northeast Industrial Cape Breton, Nova Scotia," internal report, Environment Canada, Atlantic Region, 1982, 73.

[18] G.R. Sirota, J.F. Uthe, et al, "Polycyclic Aromatic Hydrocarbons in American Lobster (*Hoarus americanus*) and Blue Mussels (*Mytilus edulis*) Collected in the Area of Sydney Harbour, Nova Scotia," *Canadian Manuscript Report of Fisheries and Aquatic Sciences*, No. 1758, June 1984.

[19] G.L. Trider and O.C. Vaidya, "An Assessment of Liquid Effluent Streams at Sydney Steel Corporation," Environment Canada, Atlantic Region, December 1980, 16, 30.

[20] M. Earle, "Sydney Steel Plant—A Chronology."

[21] Nova Scotia Department of Justice, "Sydney Tar Ponds Clean-up Project—Legal Review," draft document, released under Nova Scotia *Freedom of Information and Protection of Privacy Act* with many sections severed and suppressed, 1996. (The document itself is undated, but Nova Scotia Justice Department personnel advise it was written in June 1996.)

[22] McMullen and Smith, "Toxic Steel," citing *Mortality Atlas for Canada*, 1983.

[23] J.R. Hickman, "Health Hazards due to Coke Ovens Emissions," unpublished report, Bureau of Chemical Hazards, Health and Welfare Canada, Ottawa, April 3, 1985.

[24] J.R. Hickman, letter to Ed Norrena, Environment Canada, 30 August 1985.

[25] E. Norrena, regional director general, Atlantic Region, Environment Canada, letter to Nova Scotia deputy minister of the environment, A.H. Abbott, 1985.

[26] McMullen and Smith, "Toxic Steel," citing a 1985 report by Pierre Lavigne to the Department of Health.

[27] Nancy Robb, "Were jobs more important than health in Sydney?" *Canadian Medical Associaton Journal* 152, no. 6 (1995): 919–923.

[28] Bertrand Chau et al, "Mortality in retired coke oven plant workers," *British Journal of Industrial Medicine*, 50 (1993): 127–135.

[29] Robb Nancy, "Were jobs more important than health in Sydney?"

[30] Personal recollections. Teresa Boyd was a dear friend and well-remembered for her daily interventions on CJCB's "Talkback" program hosted by Norris Nathanson. She was one of the first people to start public demands for a cleanup of the tar ponds. (EM)

[31] "Health statistics 'not new'—coke ovens to reopen," *Cape Breton Post*, 6 November 1985.

[32] P. Donham, "No Smoke, No Baloney," *At the Centre*, Canadian Centre for Occupational Health and Safety, Ottawa, 1986.

[33] Personal notes from when I worked as Tom McMillan's senior policy adviser. (EM)

[34] S. MacLeod, "Massive cleanup of Sydney tar ponds a project whose time has finally come," *Cape Breton Post*, 16 January 1988.

[35] "The Rebirth of Muggah Creek—An Overview of the ten-year 1987–1997 Sydney Tar Ponds Clean-Up Project," Environment Canada and Nova Scotia Department of the Environment, February 1989.

Chapter Five: Murphy's Career

[1] Donnie MacPherson, "Tar Pond Tango," *New Maritimes*, March/April 1990, 16.

[2] Nova Scotia Department of Justice, "Sydney Tar Ponds Clean-Up Project—Legal Review," draft document, released under *Nova Scotia Freedom of Information and Protection of Privacy Act* with many sections severed and suppressed, 1996.

[3] "Contract handed out for tar pond clean-up," *Chronicle Herald*, 14 July 1989.

[4] Nova Scotia Department of Justice, "Sydney Tar Ponds Clean-Up Project—Legal Review."

[5] P. Donham, "Tar Ponds: Action needed now," *Halifax Daily News*, 11 October 1992.

[6] Gary Campbell, Department of Transportation and Public Works, interview with Linda Pannozzo, 8 February 1999.

[7] Nova Scotia Department of Justice, "Sydney Tar Ponds Clean-Up Project—Legal Review." 71, citing Acres's performance report.

[8] Kevin Cox, "PCBs mire waste dump cleanup—Incinerator at N.S. tar ponds not powerful enough to burn toxic compounds," *Globe and Mail*, 13 October 1992.

[9] "The taxpayer takes a bath," *Halifax Mail Star*, 5 April 1993.

[10] "Tories' tarponds buy 'underhanded'," *Daily News*, 3 April 1993.

[11] Nova Scotia Department of Justice, "Sydney Tar Ponds Clean-Up Project—Legal Review."

[12] Gary Campbell, interview with Linda Pannozzo.

[13] B. Ward, "Contaminated Sydney tar ponds passed on to Adams," *Chronicle Herald*, 9 February 1994.

[14] S. McLeod, "Tar ponds project burned cash, little else—report," *Chronicle Herald*, 1 October 1996.

[15] J. Zatzman, "Crown assumes tar ponds cleanup—Critics question reduction of incinerator testing time," *Chronicle Herald*, 8 September 1994.

[16] J. Meek, "Cleanup, coverup, dustup at Tar Ponds session," *Chronicle Herald*, 16 January 1996.

[17] M.-E. MacIntyre, "Tar Ponds Burial Slagged: Local politicians kept in dark about details," *Chronicle Herald*, 16 January 1996.

[18] T. MacDonald, "PCB testing on hold—Federal, provincial meeting to be set," *Cape Breton Post*, 11 June 1996.

[19] The letter was given to Sierra Club of Canada by Russell MacLellan.

[20] Editorial, "Plan mired in sludge," *Chronicle Herald*, 22 April 1996.

[21] M.-E. MacIntyre, "Tar ponds burial plan put on ice—Marchi expresses shock after touring the site," *Chronicle Herald*, 13 August 1996.

[22] Ibid.

[23] Sergio Marchi's wife, Laureen, shared this with me when our daughters attended the same school. As for the rest of the hotel meeting, I was in attendance and these are my own notes. (EM)

[24] M.-E. MacIntyre, "Tar ponds burial plan put on ice."

Chapter Six: JAG: Analysis Paralysis

[1] See Chapter Four for greater detail on 1983 *Mortality Atlas* data.

[2] Yang Mao et al, "Mortality in Cape Breton, Nova Scotia: 1971–1983," Special Report No. 11, Chronic Diseases in Canada, Non-Communicable Disease Division, Bureau of Epidemiology, Laboratory Centre for Disease Control, Health and Welfare Canada, Ottawa, 1 December 1985.

[3] See Chapter Four for Dr. Lavigne's comments on cancer and coke ovens.

[4] P.M. Lavigne, "A Report on the Cape Breton Risk Factor Study," Nova Scotia Department of Health, February 1987, 2.

[5] Ibid. 2.

[6] M.-E. MacIntyre, "Marchi announces cleanup funds: $1.67m given tar ponds project; goal set at 5 to 10 years," *Chronicle Herald*, 31 January 1997.

[7] Kevin Cox, "Sydney tar ponds to be probed yet again: Site still toxic after a decade of spending," *Globe and Mail*, 31 January 1997.

[8] John Campbell, "Cancer study still needed, says Ervin," *Cape Breton Post*, 28 June 1993.

[9] Canadian Press wire service, "Biologist suggests evacuation—Offers Love Canal–style solution for tar ponds," *Cape Breton Post*, 24 June 1996.

[10] T. MacDonald, "Citizens query solution: Tar pond neighbours not ready to move," *Cape Breton Post*, 25 June 1996.

[11] B. Underhill, "Ottawa may fence in tar ponds until plan is devised, Marchi says," *Chronicle Herald*, 17 September 1996.

[12] "Tar ponds circus safe," *Cape Breton Post*, 24 July 1998.

[13] Mike Britten, program co-ordinator of the Joint Action Group, interview with Linda Pannozzo, 29 January 1999.

[14] Tera Camus, "Bad soil or not, school's in: Whitney Pier parents convince board to begin classes," *Chronicle Herald*, 10 September 1996.

[15] Sharon Montgomery, "Pier residents angry," *Cape Breton Post*, 7 September 1996.

[16] Muggah Creek Watershed Joint Action Group for Environmental Clean-Up, Memorandum of Understanding Among the Government of Canada, the Government of the Province of Nova Scotia, the Cape Breton Regional Municipality, Joint Action Group for the Clean-Up of the Muggah Creek Watershed Association (JAG), signed 19 September 1998 at Sydney, Nova Scotia.

[17] S. MacInnis, "JAG expels activist: Steering committee passes code of conduct," *Cape Breton Post*, 8 April 1997.

[18] Interview with Bruno Marcocchio in *Toxic Partners*, video documentary by Neal Livingston, Black River Productions, produced in association with Vision Television, the North American Fund for Environmental Cooperation, and the Sierra Club of Canada, Fall 1998, 50 minutes.

[19] Tanya Collier, "JAG member critical of bickering," *Cape Breton Post*, 14 January 1999.

[20] Tera Camus, "Harbour sewer line stirs up debate—Move 'lesser of two evils' in Sydney tar ponds battle," *Chronicle Herald*, 28 July 1998.

Chapter Seven: Frederick Street

[1] Juanita and Rickie McKenzie, interview by authors, 27 February 1999.

[2] Debbie Ouellette, interview by authors, 27 February 1999.

[3] Kelly Toughill, "$60 million shame of `Poison City,' " *Toronto Star*, 19 July 1998.

[4] Debbie McDonald, interview by authors, 26 February 1999.

[5] Mike Britten, interview by authors, 24 February 1999.

[6] Interview with Debbie Ouellette in *Toxic Partners*, video documentary by Neal Livingston, Black River Productions, produced in association with Vision Television, the North American Fund for Environmental Cooperation, and the Sierra Club of Canada, Fall 1998, 50 minutes.

[7] Amy Smith and Tera Camus, "MacEwan downplays arsenic danger," *Chronicle Herald* May 1998.

[8] Ibid.

[9] Kelly Toughill, "$60 million shame of 'Poison City.' "

[10] Funding to hire the researcher, Dan Bunbury, was provided by the Sierra Club of Canada.

[11] Deborah Nobes, "Houses built next to toxic wasteland," *New Brunswick Telegraph-Journal* 14 September 1998.

[12] Laurel Munroe, "No immediate health risk on Frederick Street; report," *Cape Breton Post*, 23 July 1998.

[13] Staff, "Toxics and risk," *Wall Street Journal*, 3 January 1985.

[14] CANTOX, "Human Health Risk Assessment of Frederick Street Area," prepared for Nova Scotia Department of Health and Health Canada, 11 August 1998, ii.

[15] Ibid.

[16] Ibid.

[17] Ibid., unnumbered first page.

[18] R. Bertell, and R. Dixon, "Review of the Health Risk Assessment of the Frederick Street Area," conducted for the Sierra Club of Canada, by International Institute of Concern for Public Health, 28 January 1999, 37.

[19] Letter from Dr. Jeff Scott to Linda Pannozzo, 12 March 1999.

Chapter Eight: Studied to Death

[1] This was the day that the authors visited Frederick Street, Maude Barlow's first trip, and saw this pool of tar first-hand.

[2] E. May, "Action Alert: Sydney Tar Ponds," *SCAN—Sierra Club Activist News*, Fall 1998.

[3] Patty Doyle, testimony to the Select Committee on the Workers' Compensation Act, Nova Scotia Legislative Assembly, Hansard, 22 September 1998.

[4] P. Band and M. Camus, "Mortality Study of Cape Breton County and Sydney, Nova Scotia: Standardized Comparisons with Canada, 1951–1994," Environmental Health Directorate, Health Canada, September 1988.

[5] Paul Schneidereit, "Cancer risk acute in Sydney," *Chronicle Herald*, 21 October 1998.

[6] Ibid., and J.R. Guernsey, R. Dewar, S. Kirkland and S. Weerasinghe, "Cancer Incidence in Cape Breton County," Dalhousie University, 1998.

[7] Paul Schneidereit, "Cancer risk acute in Sydney."

[8] Ibid.

[9] Natalie MacLellan, "High cancer rates raise fear in Cape Breton: Dalhousie University study shows increased risk in Sydney area," *The Gazette*, 3 December 1998.

[10] John Campbell, "MacLellan calls for more study following Guernsey's cancer report," *Cape Breton Post*, 22 October 1998.

[11] Lisa Clifford, "Cancer chief urges calm for C.B.: No reason for panic, new appointee says," *Chronicle Herald*, 1 October 1998.

[12] Steve McInnis, "Firm had no work site plan," *Cape Breton Post*, 4 March 1999.

[13] Ibid.

[14] Estanislao Oziewicz, "Job-fund controversy spreads to Nova Scotia: Critics allege payoffs were used to secure cash from Ottawa," *Globe and Mail*, 26 June 1999.

[15] Ibid.

[16] Juliet O'Neill, "Pathos, politics and patronage: The anatomy of a boondoggle," *Ottawa Citizen*, 9 February 2000.

[17] Tanya Collier, "JAG calls for halt to Sobey's store project," *Cape Breton Post*, 25 March 1999.

[18] Tanya Collier, "Sobey's will continue to build store," *Cape Breton Post*, 27 March 1999.

[19] Tanya Collier, "Sobey's confident they have answer to environmental concerns," *Cape Breton Post*, 31 March 1999.

[20] Editorial, "Taking aim at cancer in Cape Breton," *Chronicle Herald*, 23 April 1999.

[21] Pierre Band and Michel Camus, "Comments on the Schabas Report," 17 May 1999.

[22] Editorial,"Taking aim at cancer in Cape Breton."

[23] Linda Dodds, "Birth Outcomes among Residents of Cape Breton County, Nova Scotia," Reproductive Care Program of Nova Scotia, 10 May 1999.

[24] H. Dolk, M. Vrijheid et al, Environmental Epidemiology Unit, Dept. Of Public Health & Policy, London School of Hygiene and Tropical Medicine (with collaborating researchers in 12 centres in five European nations), "Teratogenic Effects and Landfills," *Lancet*, 1998, 352: 423–427.

[25] Tera Camus, "Birth defects higher in Sydney—study: Doctor won't link results to toxic sites," *Chronicle Herald*, 27 May 1999.

[26] Beverley Ware, "No magic wand for toxic problem; Province won't pay to move residents," *Chronicle Herald*, 25 April 1999.

[27] Tanya Collier, "Similarities between Love Canal, Frederick Street noted," *Cape Breton Post*, 14 May 1999.

Chapter Nine: The Long Hot Toxic Summer

[1] Tera Camus, "Protester may pitch tent near premier's home—Resident from arsenic area faces hotel eviction," *Chronicle Herald* 19 June 1999.

[2] Personal notes. For much of the tent city protest, I was a participant, and conversations reported and events in this chapter are from my notes unless cited from other sources. (EM)

[3] Sharon Montgomery, "Tent pitched in protest: Whitney Pier woman camps near premier's home to make a point," *Cape Breton Post*, 21 June 1999.

[4] Ibid.

[5] "Protesters get toilet facilities," *Cape Breton Post*, 3 July 1999.

[6] Amy Smith, "MacLellan blames NDP for protest near home," *Chronicle Herald*, 26 June 1999.

[7] Ibid.

[8] Chris Connors, "JAG unable to reach consensus on incineration motion," *Cape Breton Post*, 31 August 1999.

[9] Tera Camus, "Creek scares Sydney residents," *Chronicle Herald*, 23 July 1999.

[10] Tera Camus, "One man's mission: Three people overcome as environmentalist digs into tar ponds," *Chronicle Herald*, 24 August 1999.

Chapter Ten: Leaving Home

[1] Tera Camus, "Frederick Street debris removal hits snag: C.B. strikers put stop to carting away of demolished home," *Chronicle Herald*, 30 October 1999.

[2] Tera Camus, "One of founding JAG members gets the boot: Almost half of group walk out to protest firing,"*Chronicle Herald*, 28 October 1999, 1.

Postscript: A Canadian Dilemma

[1] Clifford Lincoln, interview by authors, 28 September 1999.

[2] John Goddard, *Last Stand of the Lubicon Cree* (Vancouver: Douglas and McIntyre, 1992).

[3] Jim Bronskill and James Baxter, "Government on the hook for billions," *Ottawa Citizen*, 27 October 1999.

[4] Andrew Duffy, "Toxic waste cleanup list in the works: federal environment minister preparing national strategy," *Ottawa Citizen*, 12 October 1999.

[5] Ibid.

[6] Anne McIlroy, "Activist's fight turns terribly personal," *Globe and Mail*, 6 October 1999.

[7] Gay Abbate, "Port Hope residents near nuclear refinery demand health study," *Globe and Mail*, 8 October 1999.

[8] Anne McIlroy, "Activist's fight turns terribly personal."

[9] Warner Troyer, *No Safe Place*, (Toronto: Clarke, Irwin and Co., 1977).

[10] Donna Jacobs, "How industry beat the Environmental Protection Act," *Ottawa Citizen*, 7 September 1999.

[11] Ibid.

[12] Martin Mittelstaedt, "Ontario to receive hazardous waste—Ottawa approved transfer of 473,000 tonnes of toxins from Michigan, papers show," *Globe and Mail*, 27 September 1999.

[13] Daniel Pruzin, *International Trade Reporter*, 16 no. 27 (7 July 1999): 1124. ISSN 1523-2816 World News.

Index